Cybersecurity and Applied Mathematics

Cybersecurity and Applied Mathematics

Leigh Metcalf

William Casey

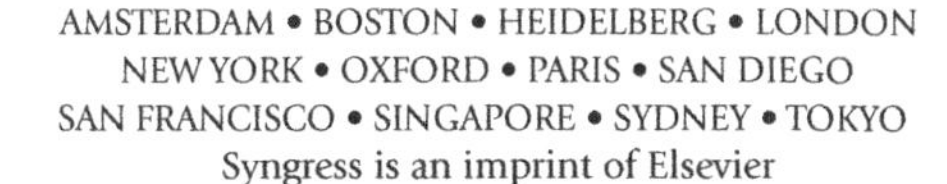
AMSTERDAM • BOSTON • HEIDELBERG • LONDON
NEW YORK • OXFORD • PARIS • SAN DIEGO
SAN FRANCISCO • SINGAPORE • SYDNEY • TOKYO
Syngress is an imprint of Elsevier

Syngress is an imprint of Elsevier
50 Hampshire Street, 5th Floor, Cambridge, MA 02139, USA

Library of Congress Cataloging-in-Publication Data
A catalog record for this book is available from the Library of Congress

British Library Cataloguing-in-Publication Data
A catalogue record for this book is available from the British Library

ISBN: 978-0-12-804452-0

For information on all Syngress publications
visit our website at https://www.elsevier.com/

Acquisition Editor: Brian Romer
Editorial Project Manager: Anna Valutkevich
Production Project Manager: Mohana Natarajan
Cover Designer: Mark Rogers

Typeset by SPi Global, India

Contents

Biography

Leigh Metcalf researches network security, game theory, formal languages, and dynamical systems. She is Editor in Chief of the Journal on Digital Threats and has a PhD in Mathematics.

William Casey is a Senior Research Member at the Carnegie Mellon University Software Engineering Institute. His work focuses on the design of scalable cyber-security within social technological systems. Casey has made contributions in the areas of cybersecurity, natural language processing, genomics, bioinformatics, and applied mathematics in academic, industry, and government settings. He has held appointments at the University of Warwick and New York University. Casey received his PhD in applied mathematics from the Courant Institute at New York University. He also holds an MS in mathematics from Southern Illinois University Carbondale, and an MA in mathematics from the University of Missouri Columbia. Casey is a member of the Association for Computing Machinery (ACM).

CHAPTER

Introduction

1

The practice of cybersecurity involves diverse data sets, including DNS, malware samples, routing data, network traffic, user interaction, and more. There is no "one size fits all" analysis scheme for this data, a new method must be created for each data set. The best methods have a mathematical basis to them.

A mathematical model of a system is an abstract description of that system that uses mathematical concepts. We want to take the systems in cybersecurity and create mathematical models of them in order to analyze the systems, make predictions of the system, derive information about the system, or other goals, depending on the system. This book is designed to give you the basic understanding of the mathematical concepts that are most relevant to designing models for cybersecurity models.

Cybersecurity is often about finding the needle in the needlestack. Finding that one bit that looks almost, but not quite like, everything else. In a network that can generate gigabytes of traffic a day, discovering that small amount of anomalous traffic that is associated with malware is a difficult proposition. Similarly, finding the one set of maliciously registered domains in the hundreds of million of domain names is not an easy process.

There are a wide variety of mathematical techniques that can be used to create methods to analyze cybersecurity data. These techniques are the underpinnings that essentially "make it work." Statistics cares about the origin of the data, how it was collected, and what assumptions you can make about the data. Mathematical techniques, such as graph theory, are developed on the structure known as a "graph," and work no matter what they are used to model. That is the beauty of math.

The point of this book is not to spend time going through proofs as to why the various mathematical techniques work, but rather to give an introduction into the areas themselves. Careful consideration was taken in the chapters to include the description of "what" things are and "how" they work, but to not overwhelm the reader with the "why." The "why" is not always relevant to understanding the "what" or "how." This book is designed for the cybersecurity analyst who wishes to create new techniques that have a secure foundation in math.

Cybersecurity and Applied Mathematics. http://dx.doi.org/10.1016/B978-0-12-804452-0.00001-4

The content is designed to cover various areas that are used in cybersecurity today, to give the reader a firm basis in understanding how they can be applied in creating new analysis methods as well as to enable the reader to achieve greater understanding of current methods. The reader is expected to have studied calculus in order to understand the concepts in the book.

CHAPTER

Metrics, similarity, and sets

2

The human eye can discern differences between two objects, but cannot necessarily quantify that difference. For example, a red apple and a green apple are obviously different, but still similar in that they are both apples. If we consider a red apple and a computer, they are obviously completely different. We can only say "similar but different" or "obviously different."

We need to quantify this difference in a reasonable way. To this end, we create a framework that standardizes the properties that a distance should satisfy. Before we cover the distance, we begin with the basics of set theory. Distances are inherently defined on a set, so the knowledge of some basics of set theory is useful. The chapter concludes with relevant examples of distances.

2.1 INTRODUCTION TO SET THEORY

We have lots of words for collections of things, a deck of cards, a flock of birds, a pod of whales, a murder of ravens, or a fleet of automobiles. The underlying theme of these words is collection of things that have a property in common; or as we say it in math, a **set** of things such as a set of birds or a set of domains. A set of domains would not contain a URL, because then we could not call it a set of domains. On the other hand, a set of domains could contain 127.0.0.1 because an IP address can act like a domain. The important notion within set theory is the idea that we can regard any collection of objects as a single thing.

We tend to use capital letters to represent sets, like A or D. In that vein, let A represent a set of Autonomous Systems represented by $A = \{\text{ASN65512}, \text{ASN65524}, \text{ASN65517}, \text{ASN65530}, \text{ASN65527}\}$. Another way to represent a set is called a Venn diagram. If A is a set, then the Venn diagram for A is in Fig. 2.1. The set is shaded because we are interested in every element within the set.

In mathematical terms, the things that make up a set are called **objects** or **members**. If an object a is contained within a set denoted by D, we write $a \in D$. If the object is not contained within D, we write $a \notin D$. Returning our set A, we know ASN65512 $\in A$ and ASN65518 $\notin A$.

Cybersecurity and Applied Mathematics. http://dx.doi.org/10.1016/B978-0-12-804452-0.00002-6

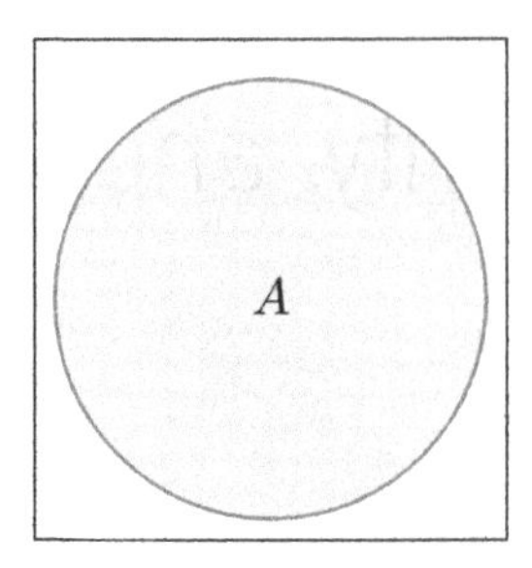

FIG. 2.1

A single set.

An important set is the **empty** set. This is the set that does not have any members at all. We represent this set as $\emptyset$ or $\{\}$. If we say that $B = \emptyset$ then we are saying that the set denoted by B contains nothing.

The number of elements in a set is called the **size**. The size of a set A is written as $|A|$. For the empty set B, $|B| = 0$ and for the set of Autonomous Systems A, $|A| = 5$.

If every element of a set is contained in another set, then the first set is a **subset** of the second set and the second set is called a **superset** of the first set. We use the $\subset$ symbol to denote a subset. If it is possible that the subset is equal to the set containing it, we use the symbol $\subseteq$. Conversely, if B is a superset of A, that is, B contains A, we can write $B \supset A$ or $B \supseteq A$.

In order to show that two set A and B are equal, we can show that $A \subset B$ and $B \subset A$. In other words, we show that every element of A is an element of B and every element of B is also an element of A. So we cannot have an element of A that is not also in B and vice versa.

We can also show that size of the subset is less than or equal to the size of the superset. If X is a subset of Y, then $|X| \leq |Y|$. It can't be more than $|Y|$, because every element of X is contained in Y so it doesn't have additional elements.

As an example, a set of domains found in malicious software is a subset of the set of all domains. So if we let $\mathcal{D}_M$ be the set of domains found in malicious software and $\mathcal{D}$ be the set of all domains, then $\mathcal{D}_M \subset \mathcal{D}$. A single element of a set is can be a subset of a set, since there is no rule that a set must contain multiple elements. So if www.example.com $\in \mathcal{D}_M$, then {www.example.com} $\subset \mathcal{D}_M$. And since the empty set contains nothing, then the empty set is contained in every set as a subset, or in other words, something contains nothing.

Example 2.1.1. A set can be considered a single object. Remember that a set is a way of considering a group of objects as a single thing. So there's no rule against having a set that contains other sets, or a set of sets. The **power set** of a set is the set of all possible subsets of that set, including the empty set and the set itself. If S is a set, then $\mathcal{P}(S)$ is the **power set** of the set. We also know that $|\mathcal{P}(S)| = 2^{|S|}$.

As an example, consider the set $S = \{a, b, c\}$. Then the power set of S is the set $\mathcal{P}(S) = \{\{a\}, \{b\}, \{c\}, \{a, b\}, \{a, c\}, \{b, c\}, \{a, b, c\}, \emptyset\}$. And we can see that that set has eight elements.

2.2 OPERATIONS ON SETS

If we start with a set, then we should have some operations defined on the sets in order to create new ones. We'll begin with elementary operations and then develop more advanced operations. The more advanced operations can be derived from the elementary operations.

2.2.1 COMPLEMENT

Suppose we have a set P and a subset Q of P. We know that every element of Q is an element of P but every element of P is not necessarily an element of Q. The elements of P that are not elements of Q are the **complement** of Q in P. It is written as Q^c. So if we have a set S of all systems in a company and a subset V of those systems that were infected by a virus, then V^c is the set of all systems in the company that were not infected. Similarly, returning to $\mathcal{D}_M$, $\mathcal{D}_M^c$ is the set of all domains that were not found in malicious software.

2.2.2 INTERSECTION

If we have two sets that have properties in common, we can find out which elements are in both sets and create a new set. This is the **intersection** of the sets. It does not make sense to find the intersection of a set of user names and a set of domains, because the two sets share no property in common. On the other hand, we can take the set of domains that our users visited and intersect it with a set of domains known to be used by malicious software. The new set is the set of domains that our users visited that are associated with malicious software.

Intersection is represented by the $\cap$ symbol. So when we intersect the sets A and B, the result is $A \cap B$. The intersection of two sets is a subset of each of the sets. That is $A \cap B \subseteq A$ and $A \cap B \subseteq B$. This is because the definition of the intersection $A \cap B$ is every element in A and B, so every element that is in $A \cap B$ must be in A. By the same argument, every element in $A \cap B$ is also in B. The Venn diagram in Fig. 2.2 is an illustration of the intersection. The shaded area between the two sets is the intersection, it represents the elements that are in both set A and set B.

Example 2.2.1. Suppose we have two sets of IP addresses. The first set will contain four IP addresses, $A = \{127.0.0.1, 192.168.5.5, 10.0.0.1, 10.5.65.129\}$ and the second will be $B = \{10.0.55.50, 10.3.5.61, 127.0.0.1, 192.168.5.5\}$. Then $A \cap B = \{127.0.0.1, 192.168.5.5\}$.

RFC1918 specifies ranges of IP addresses that should never be routed on the Internet. We will denote that by $\mathcal{R}$. We can intersect it with the data set in question

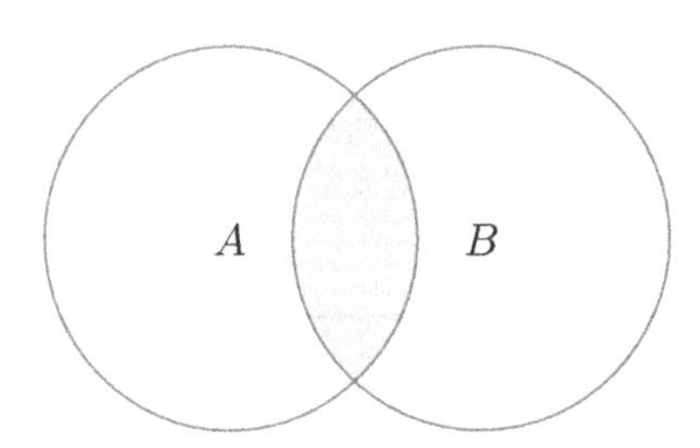

FIG. 2.2

$A \cap B$.

to determine how much of our data set should not be routed on the Internet. In this case, $\mathcal{R} \cap B = B$ and $\mathcal{R} \cap A = A$.

Note that $A \cap B = B \cap A$. Every element that is in A and B is also in B and A, so they are equivalent. This is the mathematical concept of **commutativity**.

It is possible that the intersection of two sets is the empty set. If one set is all IP addresses external to the organization and another set is all IP addresses internal to the organization, then the intersection of those two sets is empty. When this occurs, the sets are called **disjoint**. If A is a subset of B, then A^c is disjoint from A. An element can either be in A or not in A, but not both.

There is no rule that we can only intersect two sets at a time. If we have a set for each day of the week or each day of the year, we can intersect them. In fact, we can intersect any number of sets.

2.2.3 UNION

We defined an intersection of two sets A and B as the elements that are in both A and B. The union is the combination of all of the elements in A or B. If we have a set of domain names and a set of malware samples, the union of the two becomes a set of domain names and malware samples. It really does not make sense to create this set though, because a set of domains means that the entire set contains domain names. A set of domain names and malware samples does not actually specify completely what each element of the set is. Rather, it is a qualifier of what the set could contain.

Union is denoted by $A \cup B$ and the Venn diagram that illustrates it is in Fig. 2.3. You can see that both sets are shaded in, demonstrating that every element in both A or B is included in the union. The union, as in the case of the intersection, is commutative and has no restriction on the number of sets that can be combined.

If we have two sets A and B, then $A \subseteq A \cup B$ and $B \subseteq A \cup B$. This is because every element that is in A is in $A \cup B$. This is also illustrated in the Venn diagram in Fig. 2.3.

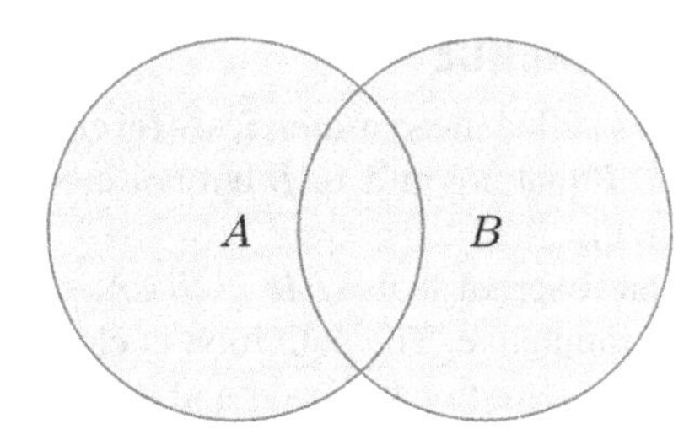

FIG. 2.3

$A \cup B$.

2.2.4 DIFFERENCE

Another combination of sets considers the elements of one set that are not in the other set. If $A = \{ASN64569, ASN64794, ASN65054, ASN65193, ASN65200\}$ and $B = \{ASN64569, ASN64575, ASN65058, ASN65200, ASN65350\}$, then the elements that are in A but not in B are $\{ASN64794, ASN65054, ASN65193\}$. On the other hand, the elements that are in B but not in A are $\{ASN64575, ASN65058, ASN65350\}$. This action is called the **difference** between the two sets and it clearly matters the order in which this action is performed. The difference between the two sets A and B is written as $A \setminus B$, while the difference between the sets B and A is written as $B \setminus A$. The Venn diagram for $A \setminus B$ is in Fig. 2.4.

We also know that if we are looking for $A \setminus B$ that this is the equivalent of finding the complement of B and intersecting that with A. The difference was described as "elements of A that are not in B." The complement was described as "not in," so that would mean we are looking at B^c. The intersection was described as "and," so the difference is then $A \cap B^c$.

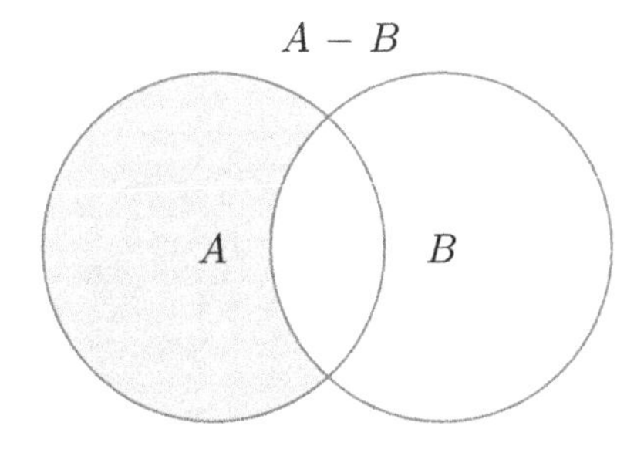

FIG. 2.4

$A \setminus B$.

2.2.5 SYMMETRIC DIFFERENCE

A variant on the difference is called the **symmetric difference**. It is comprised of the elements of two sets A and B that are in A or B but not in A and B and denoted by $A \triangle B$. The Venn diagram in Fig. 2.5.

It's clear from the Venn diagram that $A \triangle B = B \triangle A$, so unlike difference, the symmetric difference is commutative. The definition is elements that are in A or B, so we can start with $A \cup B$. The definition then says and not A and B. So the symmetric difference is as defined in Eq. (2.1):

$$A \triangle B = (A \cup B) \setminus (A \cap B) \tag{2.1}$$

2.2.6 CROSS PRODUCT

The intersection, union, subset, difference and symmetric difference of two sets creates a new set that may or may not contain all of the elements of the two sets. If we want to keep all of the elements of both sets, we can create a new set called the **cross product** or **Cartesian product** of them. Let A and B be two sets. This new set has elements that look like (a, b) where $a \in A$ and $b \in B$, so it creates pairs of every element of A combined with every element of B. The notation for this set is $A \times B$. We know that each element of A there are $|B|$ elements that are in $A \times B$, so $|A \times B| = |A||B|$.

If we're considering our set A and taking the cross product with itself, we can write $A \times A$ or A^2. So if our set is the real numbers, normally denoted by $\mathbb{R}$, we can write either $\mathbb{R} \times \mathbb{R}$ or $\mathbb{R}^2$. There's no rule that says you can only take the cross product of two sets, you can take the cross product of any number of sets. So if we take the cross product of $\mathbb{R}$ with itself n times, we write that as $\mathbb{R}^n$. Elements of this set are generally written as $(r_1, r_2, \ldots, r_n)$ and are called vectors. The set $\mathbb{R}^2$ is the standard (x, y) plane while the set $\mathbb{R}^3$ is three- dimensional space.

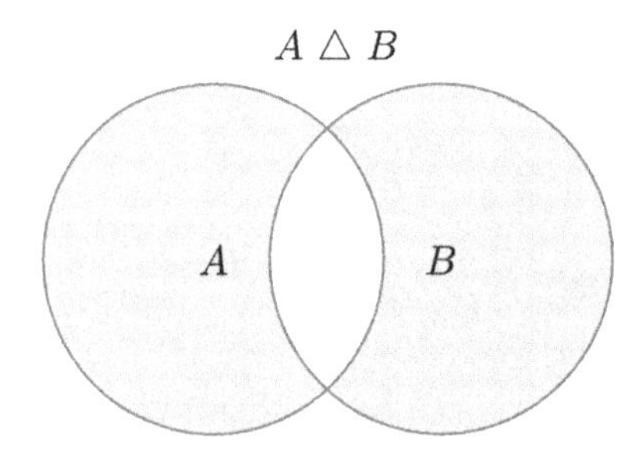

FIG. 2.5

$A \triangle B$.

2.3 SET THEORY LAWS

As we saw in the previous discussion with the definition of difference, set theory can use the operations union, intersection and complement to define new operations. They also can be used to create laws, also known as identities, that are based on the intersection and union.

We begin with the **associative** laws. The associative laws related to the order in which operations are grouped. For intersection, we know that $A \cap (B \cap C) = (A \cap B) \cap C$. This means that it does not matter what order we intersect sets in, we will get the same result. This also applies if we have four sets. We can intersect them in any order and we will get the same result if we change the order.

The associative law also applies to union. We know that $A \cup (B \cup C) = (A \cup B) \cup C$. So the order the operations are done does not affect the result. We usually write this as $A \cup B \cup C$ because the order does not matter.

Another set of laws is called the **distributive** laws. In algebra, the distributive law lets us multiply a sum by multiplying each element of the sum. In other words, $a(b + c) = ab + ac$ and $(a + b)c = ac + bc$. We have similar rules for set theory. They are $A \cup (B \cap C) = (A \cup B) \cap (B \cup C)$ and $A \cap (B \cup C) = (A \cap B) \cup (B \cap C)$.

De Morgan's laws allow us to consider the complement of the intersection of two sets. If we look at the set $(A \cap B)^c$, we can show that the this set is equivalent to the union of the complements. That is, $(A \cap B)^c = A^c \cup B^c$. Logically, this is the same as saying "not (A and B)" is the same as "not A or not B." In a similar fashion, we can show that $(A \cup B)^c = A^c \cap B^c$.

2.4 FUNCTIONS

Suppose we have two sets A and B. A **function** on the sets A and B is an operation where each element of A is paired with one, and only one, element of B. This is also called a **mapping**, where an element of A is mapped to (or associated with) an element of B. If f is our function, then we write $f : A \rightarrow B$ as the function. For an element $a \in A$ then $f(a)$ is the singular element of B that the function maps a to.

The **identity** function is a special function where each element of a set is mapped to itself. In other words, if A is our set, then $id : A \rightarrow A$ is the identity function and $id(a) = a$ for every element of A.

Example 2.4.1. Let S be a set of alphabetical strings. We define a function $f : S \rightarrow S$ by taking a string s and mapping it to the string encoded using the substitution cipher known as ROT-13. In ROT-13, each letter is replaced by the letter that is 13 letters after it in the alphabet. So a is mapped to n, b is mapped to o, m is mapped to z, and so forth. If dog $\in S$, then $f(\text{dog}) = \text{qbt}$.

The ROT-13 cipher has an interesting property to it. It is easily reversible by applying the ROT-13 cipher again. That is, $f(\text{qbt}) = \text{dog}$.

If $f : A \rightarrow B$ is a function, then the **range** of a function is all of the elements of B for which there is an element in A that is mapped to it. The **domain** of a function is the elements of A that we can map into B with f. There is no requirement that every element of B has an element of A that maps to it. It is entirely possible that a function maps to a subset of B. If every element of B has an element of A mapped to it, then the function is called **surjective** or **onto**. If it is the case that whenever function if $f(a) = f(b)$ then $a = b$, the function is called **injective** or **one-to-one**. A function that is both surjective and injective is called **bijective**.

Suppose we have two functions and we want to combine them into one. For this to work, the domain of one function must equal the range of the other function. In simpler terms, if we map from the set A to the set B, to continue the mapping we must start at B to map to the set C. So if f is the function from A to B and g is the function from B to C we can visualize it in Eq. (2.2):

$$A \xrightarrow{f} B \xrightarrow{g} C \tag{2.2}$$

The combining of two functions is called **composition**. If we are combining two functions f and g it is written as $g \circ f$ and is read right to left. If we are considering a single element, we usually write it as $g(f(a))$. This means "first operate f on a, then apply g to that result."

Example 2.4.2. Returning to Example 2.4.1, we have a function defined on $f : S \to S$ where S is a set of strings. From the definition in the example, we know that $f \circ f(s) = s$ for any element $s \in S$. So in other words, $f \circ f(\text{cat}) = \text{cat}$ which means $f \circ f = id$.

Let's consider the case where we have two functions where the domain of one is the same as the range of the other, and vice versa. In other words, $f : A \to B$ and $g : B \to A$. We know we can compose them, so we can do $f \circ g$ or $g \circ f$. If $g(f(a)) = a$, then g is called an **inverse** of f. Similarly, if for any element $b \in B$, if $f(g(b)) = b$, then f is an inverse of g. The inverse of a function f is usually written as f^{-1}. From Example 2.4.2, we know that the function defined by ROT-13 is an inverse of itself.

If we start with two sets A and B we can construct a $\{f : A \to B\}$ which is the set of all functions between the two sets. This set can also be represented by B^A.

Example 2.4.3. Let S be the set of alphabetical strings. Then S^S is the set of all possible functions that map a string in S to a new string also in S. ROT-13, defined in Example 2.4.1 is an element of this set.

2.5 METRICS

We may wish to measure the distance of elements in a set. If X is the set of objections we wish to measure, then we can define a distance function on the cross product of the set with itself mapped to the set of positive real numbers. In other words, $d : X \times X \to \mathbb{R}$. To be considered a distance function, d must satisfy a certain set of axioms. These axioms are not specific to any particular data set, rather, they are generic so that it fits every possible set. They are defined to ensure repeatability and meaningful numbers to represent the distance.

1. (Nonnegativity) $d(x, y) \geq 0$
The distance between two objects must be greater than or equal to zero. Think of this in terms of a distance between two cities. A distance of 5 miles makes sense. A distance of −5 miles does not.
2. (Coincidence) $d(x, y) = 0$ if and only if $x = y$
A distance function must be able to distinguish two objects. If the distance between two objects is 0, then they must be the same object. Otherwise, it is unable to determine the objects are different.
3. (Symmetry) $d(x, y) = d(y, x)$
A distance function must be symmetric. This means the distance between x and y must be the same as the difference between y and x.
4. (Triangle inequality) $d(x, y) + d(y, z) \geq d(x, z)$
The triangle inequality is defined in geometry. For a triangle in the plane, the sum of the length of two sides of the triangle is greater than or equal to the length of the third size. The mathematician Euclid proved this in his book Euclid's elements written in c.300 BC. By requiring this rule, we ensure that distance for a set has a reasonable geometric interpretation.

Example 2.5.1. Consider the function $d : X \times X \to \mathbb{R}^+$ by $d(x, y) = 0$ if $x = y$ and $d(x, y) = 1$ otherwise. This function is a distance function. It is non-negative and follows the coincidence requirement by definition. It is also symmetric by definition. If we have three distinct objects, then the triangle inequality also holds because 2 is greater than 1. This function is called the **discrete metric**. It is a simple example that fits the distance function requirements but in terms of examining a red apple versus a green apple the discrete metric does not show us whether those objects are closer than a red apple and a car.

2.6 DISTANCE VARIATIONS

We can vary the basic definition of a distance function and create new functions that can be used to measure difference between objects. Unfortunately, there is no definitive agreement on the terms for the new definitions. We will use terms that are often found in literature.

2.6.1 PSEUDOMETRIC

A **pseudometric** is a distance function that possibly fails the coincidence requirement. That requires that if $d(x, y) = 0$ then $x = y$. In the pseudometric, it is possible that $x \neq y$. In other words, means that the distance may fail to distinguish to objects in the original set.

Example 2.6.1. A simple example of a pseudometric can be defined using the absolute value on the (x, y) plane. If we have the points (x_1, y_1) and (x_2, y_2), then

$d((x_1, y_2), (x_2, y_2)) = |x_1 - y_1|$. We can show that this is a metric, except that the distance between the points $(1, 5)$ and $(1, 10)$ is 0, but they are obviously not equal.

2.6.2 QUASIMETRIC

If a function satisfies all of the requirements of the distance function except for symmetry, then it is a **quasimetric**. In this case, $d(x, y)$ might not be the same as $d(y, x)$. A simple illustration of this metric is a city that only has one way streets. The distance between home and the office may not be the same as the distance between the office and home, depending on the streets.

Example 2.6.2. For a practical example of a quasimetric, let W denote the set of all web pages. We will let p_i represent a single web page. Then the distance function $d(p_i, p_j)$ is the fewest number of links we have to click to get from p_i to p_j. So if we make no click, then we are on the original page, and so $d(p_i, p_i) = 0$. Similarly, if $d(p_i, p_j) = 0$, then $p_i = p_j$.

Next is the triangle inequality. If p_i, p_j and p_k are all web pages, then we can get from p_i to p_j in $d(p_i, p_j)$ clicks and from p_j to p_k in $d(p_j, p_k)$ clicks. So the greatest value that the minimum number of clicks between p_i and p_k can be is the sum $d(p_i, p_j) + d(p_j, p_k)$.

Now, we can get from p_i to p_j in $d(p_i, p_j)$ clicks, but depending on who set up the web pages and how they did it, the number of clicks between p_j and p_i may not be the same. For example, the owner of www.example.com may link to www.example.net directly, so the distance between them is one. There is no rule that owner of www.example.net has to link directly back to www.example.com. In fact, you may have to click through multiple pages to get back to www.example.com.

2.6.3 SEMIMETRIC

If all of the requirements for the distance function are true except for the triangle inequality, it is called a **semimetric**. In other words, there is no reasonable geometric realization for the metric. Network delay is commonly measured as a semimetric.

2.7 SIMILARITIES

A metric defines the distance between two objects or how far apart two objects are. If we want to measure closeness in terms of similarity, we can use another function called a **similarity measure** or **similarity coefficient**, or sometimes just a **similarity**.

The similarity function operates on the cross product of a set similar to the distance function metric. A similarity function is defined as $s : X \times X \to \mathbb{R}$. Such a function is often limited to the range $[0, 1]$ but there are similarities that return negative results. In the case of a metric we know that if $d(x, y) = 0$ then $x = y$. For a similarity function with a range of $[0, 1]$, if $s(x, y) = 1$ then $x = y$. This mean that the larger the value of the similarity function, the closer the two objects are.

Similarity functions must also be symmetric, meaning $s(x, y) = s(y, x)$. Depending on the definition of the function, there could be a variation of the triangle inequality, but a similarity function is not required to satisfy the triangle inequality axiom. As opposed to the distance function, a similarity is more vaguely defined.

If we have a similarity function where the range is $[0, 1]$, then we can at least derive a semimetric from the function. If $s : X \times X \to [0, 1]$ is the similarity function, then the semimetric will be given by $d(x, y) = 1 - s(x, y)$. If it can be shown that the triangle inequality holds for the semimetric, then we've created a distance function from the similarity. If it is the case that a similarity doesn't fulfill the coincidence requirement of a distance function, the derived function is a quasimetric.

Example 2.7.1. We'll start with the set $\mathbb{R}^n$ and two vectors $x, y \in \mathbb{R}^n$. The dot product $x \cdot y$ is an operation on the vectors that returns a single number. It is the equation $x \cdot y = \sum_{i=1}^{n} x_i y_i$. The norm of a vector x is $\|x\| = \sqrt{\sum_{i=1}^{n} x_i^2}$:

The **cosine similarity** is in Eq. (2.3).

$$CS = \frac{x \cdot y}{\|x\| \|y\|} \tag{2.3}$$

The cosine similarity is a number between 0 and 1 and is commonly used in plagiarism detection. A document is converted to a vector in $\mathbb{R}^n$ where n is the number of unique words in the documents in question. Each element of the vector is associated with a word in the document and the value is the number of times that word is found in the document in question. The cosine similarity is then computed between the two documents.

False positives may occur if both documents heavily use common words. This can skew the computation and make it appear that the two documents are related when they are not. In this case it may be necessary to consider phrases or sentences in addition to single words in the documents.

2.8 METRICS AND SIMILARITIES OF NUMBERS

In cybersecurity, we work very often with data in numerical form. for example, the amount of traffic on a network is a numerical data point. Example 2.7.1 specifies how a document can be transformed into a vector of numerical data that can be used to determine if the document is plagiarized. In this section we consider various metrics and similarities defined on numerical data.

2.8.1 L^P METRICS

The distance between two points on a straight line is the Euclidean distance. If $x = (x_1, x_2)$ and $y = (y_1, y_2)$ are two points in the plane, then the distance is given by Eq. (2.4):

$$d(x, y) = \sqrt{(x_1 - y_1)^2 + (x_2 - y_2)^2} \tag{2.4}$$

We can extend this idea to vectors in $\mathbb{R}^n$ space. Eq. (2.4) then becomes Eq. (2.5):

$$d(x, y) = \sqrt{\sum_{i=1}^{n} (x_i - y_i)^2} \tag{2.5}$$

So we know that taking the square root of an object is the same as raising it to the $\frac{1}{2}$, so the Euclidean distance has a pattern that associates squaring the difference between the elements of the vector, summing that and taking the square root of the result. We can make Eq. (2.5) more generic by replacing the 2 with a p and we get Eq. (2.6):

$$d_p(x, y) = \left(\sum_{i=1}^{n} (x_i - y_i)^p \right)^{1/p} \tag{2.6}$$

If $p < 1$, let's look at the three points $x = (0, 0)$, $y = (0, 1)$ and $z = (1, 1)$. We want the triangle inequality to hold, that is, $d_p(x, z) \leq d_p(x, y) + d_p(y, z)$. The distance between x and y is 1 as is the distance between y and z. However, the distance between x and z is $2^{\frac{1}{p}}$. Since $p < 1$, then $\frac{1}{p} > 1$, which means that $d_p(x, z) > d_p(x, y) + d_p(y, z)$. So the triangle inequality doesn't hold for $p < 1$. For $p \geq 1$ it does hold, and this metric is called **Minkowski** distance or the L^p metric.

If $p = 2$, it's the standard Euclidean distance. For $p = 1$, it is called the **taxicab** or **Manhattan** distance. The formula for it is in Eq. (2.7):

$$d_1(x, y) = \sum_{i=1}^{n} |x_i - y_i| \tag{2.7}$$

An illustration of the taxicab metric is in Fig. 2.6. From this picture it's clear why it's called the taxicab metric. If you drive from the point $(1, 1)$ to the point $(5, 5)$ in by city blocks, you could follow the route displayed in that picture.

We can take the limit of p to infinity, that is, $\lim_{p \to \infty} d_p(x, y)$. The function becomes $d_\infty(x, y) = \max_{1 \leq i \leq n} |x_i - y_i|$, also known as the L^∞ metric or the Chebyshev metric. So the distance between the two points is the furthest distance between the coordinates in the vector. If our two points are $(1, 1)$ and $(2, 1)$, then the L^∞ distance between them is 1.

Another way to consider the L^p metric is to consider the circles drawn by the metric. In the standard Euclidean metric, we know that the circle centered at $(0, 0)$ is all points where $d_2(x, 0) + d_2(y, 0) = 1$. For the general Minkowski metric, the circle centered at $(0, 0)$ is all points where $d_p(x, 0) + d_p(y, 0) = 1$. For $p = 2$, this produces the standard unit circle. For p of other values, it doesn't. Consider Fig. 2.7.

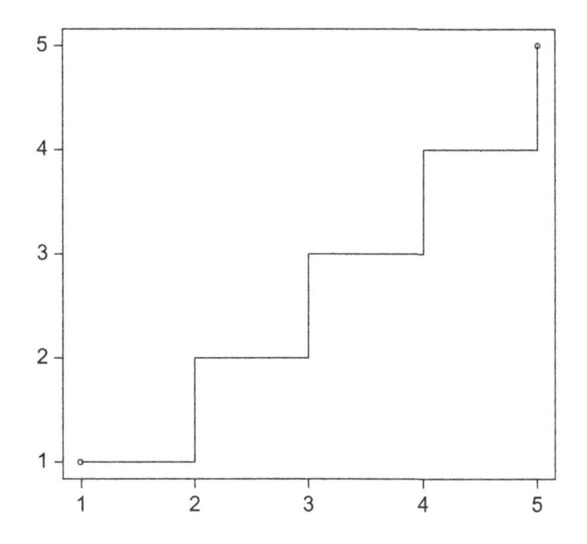

FIG. 2.6

Taxicab distance.

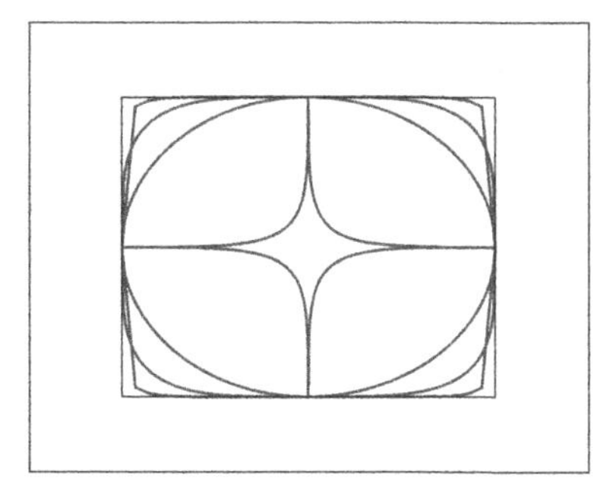

FIG. 2.7

Circles for Minkowski metrics.

The concave figure is the circle when $p = 1/3$. As p increases, the circles go from concave figures to convex, and when $p = \infty$, the circle becomes a square of radius 2.

The weighted Minkowski metric is a variation of the metric that allows us to assign weights to each element of the vector. We do this by creating a vector of weights in $\mathbb{R}^n$, denoted by $(w_1, w_2, \ldots, w_n)$. The weighted Minkowski metric is then in Eq. (2.8):

$$d_p(x, y) = \left(\sum_{i=1}^{n} w_i (x_i - y_i)^p \right)^{1/p} \tag{2.8}$$

2.8.2 GAUSSIAN KERNEL

Machine learning provides us with another similarity that is based on the Euclidean metric, so it is defined on elements of $\mathbb{R}^n$. Let $d(x, y)$ denote the Euclidean metric between two vectors x and y. The similarity is called the **Gaussian Kernel** and is in Eq. (2.9):

$$k(x, y) = \mathrm{e}^{-d(x,y)^2} \tag{2.9}$$

If $x = y$, then $d(x, y) = 0$, so $k(x, y) = 1$. From the negative sign on the exponent, we know that the values of $k(x, y)$ are between 0 and 1, and are 1 only if the two values are equal.

2.9 METRICS AND SIMILARITIES OF STRINGS

In addition to numerical data sets, strings show up as well in Cybersecurity data sets. In Example 2.7.1, we discussed translating the string data into vectors of numbers, but it is possible to define metrics and similarities directly on the strings themselves. This section will introduce common ones.

2.9.1 LEVENSHTEIN DISTANCE

In this section we define a string as an ordered collection of characters. So "xyz" is an example of a string and so is "123abc" and also "The quick red fox jumped over the lazy dog." There are three operations we can execute on a string, they are:

1. Replacement
   A character in a string can be replaced by a different character. So for example, "cat" would become "cut" if we replaced the middle character "a" in cat with a "u."
2. Insertion
   In the case of insertion, we add a new character to a string. If we insert an "o" into the word "flat," we create "float."
3. Deletion
   Deletion is the removal of a character from a string. For example, removing the first "e" from "create" will give us "crate."

Given two strings, we can transform one into the other using the above operations. If we start with "dog," then to get the string "cat" we substitute each of the three letters in "dog." If we start with "computer," then we need to delete the "m" and the "p" and insert an "n" to get "counter." The **edit distance** or **Levenshtein distance** between two strings is the minimum number of operations to convert one string into another string.

If the edit distance between two strings is 0, then the strings are equal since there's nothing to change from one to get the other. The distance is symmetric, because if we insert a character to get from one string to another, then in the reverse we delete. Substitution would have the same number of instructions both from the first string to the next second and vice versa. The question becomes the triangle inequality. If we have three strings, then the first is transformed into the second by n moves and the second is transformed to the third by m moves. So the minimum number of moves to transform from the first to the third is no bigger than $n + m$ moves.

Levenshtein distance can be used in spell checkers. If you typed the word "comuter," it has an edit distance of 1 to "computer," so a spell checker would suggest that word as the correct word. It could also suggest "commuter," since that also has an edit distance of 1.

2.9.2 HAMMING DISTANCE

Consider the set of all strings of a length n, where n is an integer. The **Hamming distance** between two strings is the number of places in which the two strings differ. For example, suppose $n = 4$ and we consider the words "jazz" and "fizz." The Hamming distance between these two strings is 2.

If the Hamming distance between two strings is 0, then they must be the same. It's also clearly symmetric, since the number of places between the first string and the second string is the same as between the second string and the first string. It takes quite a bit of an argument to prove that the triangle inequality holds, so we will omit that in this chapter. So the Hamming distance is a metric on a set of strings of a predefined length.

Since the Hamming distance is only defined on strings of the same length, if you have a varied data set the Levenshtein distance is a better choice.

2.10 METRICS AND SIMILARITIES OF SETS OF SETS

In Example 2.1.1, we defined a set of sets, where we consider a set of things as a single object and then create a set of them. As a metric is defined on a set, we can define a metric or similarity on our sets of sets. This section will discuss some of the more common variations.

2.10.1 JACCARD INDEX

We'll start with two sets A and B. We want to determine how similar the two sets are, so let's consider what we can measure that shows commonality between the sets. If we consider the size intersection of the sets, that shows us how many objects are in common between the two sets. We'll take the ratio of that with the number of objects that are in the two sets, and this will give us the similarity called the **Jaccard index** or the **Jaccard coefficient**. In precise terms, the Jaccard index is in Eq. (2.10):

$$J(A, B) = \frac{|A \cap B|}{|A \cup B|} \tag{2.10}$$

If the Jaccard index between two sets is 1, then the two sets have the same number of elements in the intersection as the union, and we can show that $A \cap B = A \cup B$. So every element in A and B is in A or B, so $A = B$.

Suppose we have two sets of domains, one that denotes domains that are known to be malicious called M and one that denotes the domains accessed at a corporation, call it C. The Jaccard index between the two sets gives us an idea of how similar the two sets are. If the sets are very similar, then the users at that corporation have been accessing quite a few malicious sites.

The Jaccard index can also be used on strings. For each string, define a set contains the characters in a string, so the string "cat" becomes $\{c, a, t\}$. If we have two strings "catbird" and "cat," then the numerator is 3 and the denominator is 7, which gives us a Jaccard index of $\frac{3}{7}$. Note that when the Jaccard index is applied to strings it doesn't consider the order of the characters in the strings. It merely considers the presence of the characters.

Example 2.10.1. Let $(0, 0, 0, 0)$ and $(0.1, 0.1, 0.1, 0.1)$ be elements of the set $\mathbb{R}^4$. Then the Euclidean distance of the two elements is 0.2. However, if we consider the Jaccard index, the similarity is 0, as the two vectors have no elements in common. So the fact that the Euclidean distance is small means that they are close together by using that metric, but the similarity given by the Jaccard index implies that they are very different.

2.10.2 TANIMOTO DISTANCE

In Section 2.7 we discussed how a similarity on $[0, 1]$ could be transformed to a quasimetric or even a metric by taking one minus the similarity. The **Tanimoto metric** is a good example of this. The Tanimoto metric, also known as the **Jaccard distance** is given in Eq. (2.11):

$$d_J(A, B) = 1 - J(A, B) \tag{2.11}$$

It can be shown that the distance is a metric. Symmetry can be shown as $A \cap B = B \cap A$ and $A \cup B = B \cup A$, the main question becomes the triangle inequality. This is a more complex argument and is omitted from this chapter to keep the focus on remaining topics.

2.10.3 OVERLAP COEFFICIENT

The **overlap coefficient** is a variation of the Jaccard index, as in Eq. (2.12) for two sets A and B:

$$O(A, B) = \frac{|A \cap B|}{\min(|A|, |B|)} \tag{2.12}$$

If the two sets A and B are of disparate size, then the value $|A \cup B|$ can overwhelm the numerator $|A \cap B$, negating the contribution of the smaller set. By choosing the minimum value of A and B, we're allowing the fraction to determine the percentage of the smaller set that is common with both sets. The coefficient is a similarity, not a metric. If the value of the similarity is one, that does mean that the two sets are equal. It means that one is a subset of the other, but it does not definitively demonstrate equality though.

2.10.4 HAUSDORFF METRIC

The **Hausdorff metric** between two sets assumes that we have a metric defined between elements of the sets. For example, if the elements of the sets are strings, we have the Jaccard distance. Similarly, if the elements of the sets are vectors, we have the L^p metric. To keep things simple, we will assume our sets are finite and discrete. In other words, we can enumerate each element of the set and there are only a finite number of them.

The metric is then the minimum distance between all of the elements in the two sets. It follows that this is a metric because it is based on a metric.

2.10.5 KENDALL'S TAU

Suppose we have a list that ranks objects. A simple example is a list that ranks the top 20 tennis players in the world. If a second organization ranks the same players, but in a different order, we would like a metric that tells us how different the two rankings are. The sorting algorithm bubble sort can take these two lists and sort one so that it looks like the other. We can then count the number of operations that this sort would take and this is called the **Kendall's tau**. This can be shown to be the minimum number of operations it takes to turn one list into another by using this sort.

We begin with two lists labeled Ł$_1$ and Ł$_2$. If x is an element in the list Ł$_1$ then x_i is the ranking of x. Similarly, if y is an element of the list Ł$_2$ then y_j is the ranking of y. We pair the elements of both lists as (x_i, y_i). Two pairs (x_i, y_i) and (x_j, y_j) are **concordant** if $x_i > x_j$ and $y_i > y_j$ or $x_i < x_j$ and $y_i < y_j$. The pairs are **discordant** if the ranks disagree, that is, $x_i > x_j$ and $y_i < y_j$ or $x_i < x_j$ and $y_j > y_i$. If the ranks agree, the pair is neither concordant or discordant.

Kendall's tau is then given in Eq. (2.13):

$$\tau(L_1, L_2) = \frac{\text{Number of concordant pairs} - \text{Number of discordant pairs}}{\frac{1}{2}n(n-1)} \tag{2.13}$$

If the two lists are equal, then $\tau(L_1, L_2) = 0$. It's also symmetric. It counts the minimum number of operations needed to turn one list into another, then normalizes that value. If we transform the first list into the second by j moves and the second into the third by k moves, then the transformation of the first into the third can be no more than $j + k$ moves. Therefore, Kendall's tau is a metric.

2.11 MAHALANOBIS DISTANCE

Often in data analysis we have a collection of related numerical data that is of completely different scale. For a simple example, suppose we have a collection of malware samples and we're considering two of each sample, the size of the malware in bytes and the number of files that it drops. We would expect a wider range of variability for possible malware sizes variable than the amount of variability for the number of files that malware drops. For example, we may have a piece of malware that is 49,779 bytes and drops two files and another piece of malware that drops one file which is 75,420 bytes. A standard Euclidean distance comparing these samples the wide range in variation of the size variable versus the smaller variation in number of files dropped would be overwhelmed by the size variable. If we consider the distance of a single point to a set of points, then the Hausdorff distance will also not give a meaningful result.

Instead of considering the items point by point, we look at them as cloud of points and consider how they're distributed in $\mathbb{R}^p$. Fig. 2.8 is an example of that in $\mathbb{R}^2$. Then to find if an element is close to this set, we compute the **Mahalanobis** distance between the point and the set.

We will call our set of points X and let n be the number of points in each set. Each element of X will be given as x_i. Since $x_i \in \mathbb{R}^p$, then we can look at the x_{i_j} as the jth member of the vector x_i. The first step is to compute the mean of the set, denoted $\bar{X}$. Since every element of X is a member of $\mathbb{R}^p$, then the mean of the set is a member of $\mathbb{R}^p$, which also means we can consider the jth element of $\bar{X}$ as $\bar{X}_j$. The second step is computing the **covariance matrix** of X. The i, j cell of the matrix is computed by the formula in Eq. (2.14):

$$\mathrm{cov}(x_j, x_k) = \frac{1}{n}\sum_{i=1}^{n}(x_{i_j} - \bar{X}_j)(x_{i_j} - \bar{X}_k) \tag{2.14}$$

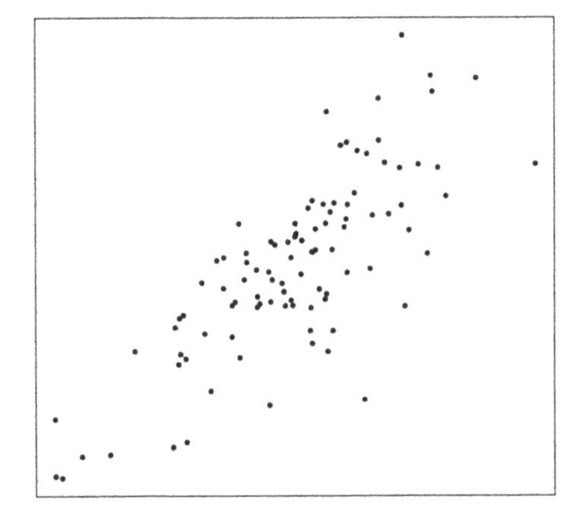

FIG. 2.8

A point cloud.

We represent the covariance matrix by Σ. Eq. (2.15) is the Mahalanobis distance between the vector x and the set of observations X:

$$d(X,x) = \sqrt{(x-\bar{X})^T \Sigma^{-1} (x-\bar{X})} \tag{2.15}$$

2.12 INTERNET METRICS

In the previous three sections we've discussed methods to determine distance and similarity on three types of data. In all three cases, the metrics and similarities are mostly independent of the data that is measured. Sometimes, the comparison of two objects should be affected by the data in the set. In this section, we will discuss Internet specific metrics. These are examples of metrics that are created by examining the data set directly.

2.12.1 GREAT CIRCLE DISTANCE

The distance between two cities can be measured in more than one way. One method to consider is the distance traveled on roads between the cities. However, that assumes you can drive between the cities. It isn't possible to drive between Sydney, Australia and Los Angeles, California. In this case, we could measure by the distance a plane travels. The danger of using the method of travel is that you have to consider every method you could use between cities, which doesn't lead to a standard metric. Instead, we will use a distance called a **great circle distance** between the latitude and longitude of our two locations on the globe.

The great circle distance is the shortest distance between two points on a sphere. In the plane, the shortest distance between two points is a straight line. A sphere doesn't have any straight lines due to the curvature of the sphere. The great circle distance takes that into account to determine the distance.

Suppose we have the latitude and longitude of two locations, $\phi_1\lambda_1$ and $\phi_2\lambda_2$. We also need the radius of the earth, which is approximately $r = 6371$ km. We'll let $\Delta\phi = |\phi_1 - \phi_2|$ and $\Delta\lambda = |\lambda_1 - \lambda_2|$. Then the Haversine formula is given in Eq. (2.16):

$$\Delta\sigma = 2 \arcsin \sqrt{\sin^2\left(\frac{\Delta\phi}{2}\right) + \cos\phi_1 \cos\phi_2 \sin^2\left(\frac{\Delta\lambda}{2}\right)} \tag{2.16}$$

The distance between the two locations is given by $d(\phi_1\lambda_1, \phi_2\lambda_2) = r\Delta\sigma$.

2.12.2 HOP DISTANCE

When a packet leaves a computer system, it traverses through several other devices before it reaches it's destination. These devices can be routers, traffic shapers, load balancers, or even other systems before it reaches it's final destination. Each device on the path from source to destination is called a **hop**. For example, the software *traceroute* displays each hop from source to destination.

The number of hops between a source and destination is called the **hop count**. It isn't a particularly reliable metric as the number of hops may change from day to day. The routing protocol RIP uses the hop count as part of its calculations.

The routing protocol BGP groups devices under the same administrative control into one entity, called an Autonomous System. Rather than counting the number of hops between a source and destination, we can count the number of Autonomous Systems between the two systems. In BGP, there may be more than one path between two hosts. As a rule, we use the fewest number of Autonomous Systems between two hosts as the distance. This is a metric.

BGP administrators use weighted hop distance in order to force particular routing choices. The distance is then the number of hops combined with the weight at each hop. The BGP operator would then choose the path with the lowest weighted distance as the best path to the destination.

2.12.3 KEYWORD DISTANCE

Web pages are often indexed by keywords as part of the search engine optimization process. Similarly, academic papers are also often indexed by keywords to facilitate searching. The keywords may have similar meanings, so we want to measure the similarity if the keywords based on the documents they index.

If we have a keyword k_1 and another k_2, we will let K_1 be the set of documents indexed by k_1 and K_2 be the set of documents indexed by k_2. Then the **semantic proximity** or **keyword distance** of k_1 and k_2 is in Eq. (2.17):

$$S_P(k_1, k_2) = \frac{|K_1 \cap K_2|}{|K_1 \cup K_2|} \tag{2.17}$$

This is the Jaccard index from Eq. (2.10). This section demonstrates an application of the index where the data is in a specific state and allows us to determine the distance between two words, rather than between two sets.

CHAPTER

Probability models

3

A basic course on probability contains a lot of problems concerning boxes containing two colors, say red and white, and determining the probability of picking one kind, that is, picking a red ball versus a white ball. These experiments can make it seem like probability is for nothing more than parlor tricks, such as "what is the probability of pulling an ace out of a deck of cards?" or "if I guess all of the answers on a multiple choice test, what is the probability that I will pass?" However, probability has much richer applications than these simple tasks. What if instead of a box containing red and white balls, we have a constant flow of red and white balls and we know that 30% of the flow is red and 70% of the flow is white. What is the probability of choosing a red ball out of this flow? Or to put it in cybersecurity terms, if I have a constant flow of packets with 90% of them "good" and 10% labeled as "bad," what is the probability of choosing a good packet from the flow?

A mathematical model is a way of using a mathematical concept to describe an event. For a simple example, suppose we have a token ring network. We can model that network by using a circle to denote the ring around which the token travels. Probability is a way of modeling real events, like the amount of time it takes for the token to travel from one host to the next. It gives us a language to describe events mathematically so we can analyze or make predictions about the events. If, for example, would like to be able to predict whether or not a traffic pattern from a host indicates a virus on that host, it is beneficial to have a framework that allows us to make these predictions. Probability is that framework, combined with statistics.

This chapter is designed to discuss the origins and applications of random variables and provide an overview of probably and the relationship between probability, random variables and statistics.

3.1 BASIC PROBABILITY REVIEW

When an event happens, it can be that something else caused that event. For example, when we switch the computer off, it turns off. The computer turns off because we flipped the switch. Alternatively, events may have no direct case at all. For example,

Cybersecurity and Applied Mathematics. http://dx.doi.org/10.1016/B978-0-12-804452-0.00003-8

a card may be selected from a deck but it may have been drawn at random from that deck. Randomness can also be affected by perception. What is random to one person, such as "my email arrived late" is not to another, who knows that the connection to the ISP was down for an hour.

To be precise, we define a **random event** as a visible or large effect with an invisible, small, or nonexistent cause. The result of flipping a coin is a random event, as is the drawing of a card from a deck of shuffled cards, or the arrival times for email.

Let us consider the outcomes of these random events by performing a **random experiment**. Random experiments are experiments that can be repeated any number of times with the same relevant conditions. They include:

- Count the number of packets that traverse a network in a 30 min interval.
- Measure the length of time in a HTTP session.
- Flip a coin.
- Count IP addresses blocked by a firewall in 1 h.

In some cases, conditions will not remain exactly the same forever. For example, when counting the packets, the state of a network could change completely in 30 minutes time. However, we are usually able to assume that they stay essentially the same for our intents and purposes. When flipping a coin, the humidity, the wind, or even the person flipping the coin could affect the outcome of the flip. If we spend all of our time considering even the tiniest of changes in environment, we will never find the time to actually analyze the event. In short, we pretend that even though the environment of the experiment can change, it will not affect the experiment drastically. If it can, then we need to reconsider the experiment.

The set of possible results for a random experiment is called the **sample space**. Common labels for the sample space are S or Ω, and we will use Ω. The elements of the sample space are **outcomes** of the random experiment and the sample space is the then the set of all possible outcomes.

If we consider the example where we are flipping a coin, the sample space is $\{H, T\}$ for head and tails. On the other hand, if we count the IP addresses blocked by a firewall in 1 h, the result can be any positive integer. In this case, our sample space is $\Omega = \mathbb{Z}^+$. For a third example, suppose we are choosing a random card from a deck. Our sample space is every card in the deck.

Example 3.1.1. We ask a computer for a random integer between 0 and 5 as our random experiment. The sample space is then $\Omega = \{0, 1, 2, 3, 4, 5\}$. Technically speaking, it could contain $\Omega = \{0, 1, 2, 3, 4, 5, S, M\}$ where S stands for the computer shutting down in the middle of the experiment because of a hardware fault and the M stands means a meteor hit and took out the power source. However, a sample space should only contain events that would reasonably happen. The hardware fault and the meteor strike are not reasonable outcomes, so they should not be included.

If our random experiment is flipping coins, then the events that could reasonably happen are $\{H, T\}$, not heads, tails, lands on an edge, hovers in the air, and is eaten by a passing bird. We need to careful consider our sample space and ensure that it only contains the reasonable events.

Now suppose we ask our computer for two random integers between 0 and 5. Our sample space has grown quite a bit and contains 36 outcomes. They all have the form i, j where i and j are numbers between 0 and 5.

While a single element of our sample space is an outcome, a subset of the sample space is an **event**. We know from set theory that $\emptyset \subset \Omega$ and since the empty set contains nothing, then the event that the empty represents means nothing happened as a result of the experiment. However, something always happens when we have a random experiment and there is always an outcome. So $\emptyset$ is sometimes referred to as an impossible event.

3.1.1 LANGUAGE AND AXIOMS OF PROBABILITY

Probability is a method of quantifying numerically exactly how likely we think it is that a certain event will occur. Without the precision of numbers, we could say "we think it will happen," "we really think it will happen" or "maybe it will happen" and order those results, but how much more likely is "we think it will happen" as compared to "we really think it will happen?" We measure this value with a number between 0 and 1. If the value is closer to 1 it is more likely to happen and if the value is closer to 0 it is less likely to happen.

The precise definition of the **probability** of an event is a function $P : \mathcal{P}(\Omega) \to [0, 1]$. That is, a function from the power set of the sample space to the interval $[0, 1]$. The properties of P include:

1. $P(\emptyset) = 0$.
2. $P(\Omega) = 1$.
3. For disjoint events A and B, that is, $A \cap B = 0$, $P(A \cup B) = P(A) + P(B)$.

Recall that $\emptyset$ is considered the impossible event, that is, the event that could never happen. So $P(\emptyset) = 0$, which tells us that when the probability of an event is 0, it could not happen. Similarly, consider $P(\Omega)$. As the value is 1, this means that the event Ω must have occurred. So for some $A \in \Omega$ if $P(A) = 1$, then A is guaranteed to occur.

Example 3.1.2. Suppose we are flipping a coin, but instead of a fair coin, we are flipping a two headed coin. Then the event $\{T\}$ cannot happen, so $P(\{T\}) = 0$. Similarly, $P(\{H\}) = 1$, as heads will always be the result.

More properties of the probability function include:

4. If $A \subseteq B$ then $P(B \setminus A) = P(B) - P(A)$.
5. $P(A^c) = 1 - P(A)$.
6. If $A_1 \cap A_2 \cap \ldots A_n = \emptyset$ then $P(\cup_{i=1}^{n} A_i) = \sum_{i=1}^{n} P(A_i)$.
7. If $A \subseteq B$ then $P(A) \leq P(B)$.

Let us briefly consider $P(A^c)$. We know that $A^c \cap A = \emptyset$ and that $A^c \cup A = \Omega$. So we say that $P(\Omega) = P(A) + P(A^c)$. However, we also know that $P(\Omega) = 1$, so $P(A) + P(A^c) = 1$ which gives us the property $P(A^c) = 1 - P(A)$. We also know that $P(A) = 1 - P(A^c)$. This discussion demonstrates how some of the properties of probability can be derived from other properties. The first three properties of probability are the most important, all others can be derived from those.

The elements of the sample space are outcomes and probability is defined on events which are subsets of the sample space. So technically speaking, we should be writing $P(\{\omega\})$ when we are considering the probability of a single event as we did in Example 3.1.2. However, we tend to abuse notation and write this as $P(\omega)$ when we are considering the probability of a single event.

Using this notation, we also know that if $A = \{\omega_1, \omega_2, \ldots, \omega_n\}$ then $P(A) = \sum_{i=1}^{n} P(\omega_i)$. This is an application of the sixth property of probability, since $\{\omega_i\} \cap \{\omega_j\} = \emptyset$ for all $\omega_i, \omega_j \in A$.

Example 3.1.3. We ask a computer for a random integer between 0 and 5 and assume that the probability of getting any integer is equal, that is, $P(i) = \frac{1}{6}$.

What is the probability that the integer is an odd number? The outcomes in our sample space that fit this requirement are 1, 3 and 5. So the probability of getting an odd number are $P(1) + P(3) + P(5) = \frac{3}{6}$.

Example 3.1.4. For our random integers, what is the probability that the integer is greater than 1? That is, we want to know $P(x > 1)$. There are two ways to solve this problem. One is to determine the events in the sample space that fit this criteria and to sum the probabilities. The other is to consider what events in the sample space are in the complement of $x > 1$ which is $x \leq 1$. The events in the complement of the space are 0 and 1, so the probability of the event $P(x > 1) = 1 - (P(0) + P(1)) = \frac{4}{6}$.

3.1.2 COMBINATORICS AKA PARLOR TRICKS

We have discussed properties of probability, let us now turn to computing probability. We start with N events, with equally likely outcomes. For example, when flipping a coin, it is equally likely we will get a head or a tails. The probability of an outcome ω in this situation is $P(\omega) = \frac{1}{N}$. We can extend this idea to the probability of an event $A \subseteq \Omega$. A has $|A|$ events, and so $P(A) = \sum_{i=1}^{|A|} P(\omega)$. Therefore, $P(A) = \frac{|A|}{N}$.

Example 3.1.5. In Example 3.1.1 we said that if we ask the computer for two random integers between 0 and 5 then the sample space contains i, j where i and j are integers between 0 and 5 and that $|\Omega| = 36$. We also assume that there is an equal probability of getting any integer from the computer.

What is the probability that the sum of the two integers is 1? There are exactly two elements in this sample space that fit that condition, 0, 1 and 1, 0. Therefore, the probability is $\frac{2}{36}$.

Alternatively, we could list the elements as i, j where i is the result of one query and j is the result of the other. In this case, we are removing order from the list. In this case, there are 21 elements of this sample space, which we will denote as Ω'. The probability that the sum of two elements having the sum of 1 in this case is $\frac{1}{21}$.

Except, that is not the case. There is only one query that would give us 0, 0 while there are two queries that would give us 0, 1. It is a valid sample space for our example, but it is not one where the elements have equally likely outcomes. If we assume equally likely outcomes, we need to be able to correctly determine the size of the sample space. This example demonstrated that if we count the elements incorrectly, we can get the wrong probabilities.

Let us turn to considering the ways of counting the number of elements in our sample space. Suppose random event 1 can be performed in n_1 ways, random event 2 can be performed in n_2 ways and random event 3 can be performed in n_3 ways. The total number of ways all three events can be performed is $n_1 n_2 n_3$.

Example 3.1.6. Suppose we require four digit passcodes to enter an area. To calculate the number of possible passcodes, we begin by considering the number of choices available for the first digit. There are 10 possible outcomes for this digit. Similarly, the remaining digits have the same number of possible outcomes. Therefore, the number of possible passcodes that are possible is $10 \cdot 10 \cdot 10 \cdot 10 = 10^4$. in other words, there are 10,000 possible codes. This makes sense, since if we rephrase the question, we can enter any number between 0000 and 9999, of which there are 10,000.

For another example, suppose our event is to ping a system and the out come is either U or D for up or down. How many different outcomes are possible if we ping it five times? The answer is 2^5 since each event has two outcomes and we perform it five times.

For a different method of counting, suppose we have four people, Alice, Bob, Carol and Dave. We will denote them by A, B, C and D. We sit them on a long bench, how many different ways can we order them? ABCD is a valid choice, as is CADB or even DBCA. So, for the first position we have four choices. In the second, we have three, the third, two and just one choice for the last position. Therefore, the number of possible ways we can order them is $4 \cdot 3 \cdot 2 \cdot 1 = 24$.

This creates a **permutation** of the set of four people. In general terms, a permutation is when we choose an ordered subset of size k from a set of size n, where $k \leq n$. The notation for a permutation is $(n)_k$ and the formula for the number of elements in a permutation is in Eq. (3.1). We can think of permutations of a set of elements as "the number of ways we can order these elements into a list":

$$(n)_k = n(n-1)(n-2)\cdots(n-k+1) \tag{3.1}$$

The **factorial** of a number n is the number multiplied by every integer less than the number that is greater than or equal to 1. We write this as $n!$ More precisely, $n! = n(n-1)\cdots 1$. For example, $5! = 120$. We define $0! = 1$.

Returning to our discussion of counting, permutations are concerned with the order of elements in a set. A **combination** is an unordered subset of a set. The subset has a size of $k \leq n$ and is denoted by $\binom{n}{k}$, which is read is n choose k. The formula for a combination is given in Eq. (3.2):

$$\binom{n}{k} = \frac{(n)_k}{k!} = \frac{n!}{k!(n-k)!} \tag{3.2}$$

While a permutation was concerned with ordering the element of a set, a combination is concerned with "choosing any elements" from the set. The restraints on a combination are less than a permutation.

Example 3.1.7. Suppose we have a reserve set of 20 servers. How many different ways can we choose 4 servers from this set? The answer is $\frac{20!}{(4!16!)} = 4845$. We can then use this sample space size to determine the probability of choosing a bad server in the set if we know that two of the servers are bad in the set and it is equally likely when we choose a server that we will get a bad server or a good server.

To reiterate the point made in Example 3.1.5, we should always be aware of exactly what our sample space is and to ensure that it describes our random experiments appropriately.

3.1.3 JOINT AND CONDITIONAL PROBABILITY

Suppose we have a deck of cards. We draw one card from the deck, what is the probability that it is a queen? Since there are four queens in a standard deck of 52 cards, so the probability of a queen is $P(\text{queen}) = \frac{4}{52}$. If we ask what is the probability that the card is a queen and is red? The answer is determined by the fact that there are two cards in the deck that are red and queens, so the answer is $P(\text{queen and red}) = \frac{2}{52}$.

We can rephrase the question in terms of events. The first event is $Q = \{\text{queens in the deck}\}$ and the second event is $R = \{\text{red cards in the deck}\}$. So the probability of a queen and red can be written as the intersection of the two events, or $P(Q \cap R)$. This is called **joint** probability.

Suppose we have 20 servers in a facility. All 20 are web servers, but only two of them allow the uploading of files. If we choose a server at random, what is the probability that we find a server that allows us to upload a file? It is straightforward without doing any calculations that the answer is $\frac{2}{20}$. Given that we found a server that will allow us to upload a file on our first event, what is the probability that we will find the other server that allows us to upload a file on our second event? We have changed the sample space in this case by removing one server in our first event. So the probability of finding a server on our second event is $\frac{1}{19}$ since there are now 19 servers in our sample space.

This is called **conditional** probability. The probability of a second event is influenced by the knowledge of the first event. If our first event is B and our second is A, then we write conditional probability as $P(A|B)$. This is read as the probability of A given that B occurred. Eq. (3.3) is the formula for conditional probability:

$$P(A|B) = \frac{P(A \cap B)}{P(B)} \tag{3.3}$$

Clearly conditional probability is related to joint probability. The rules of conditional probability are:

1. $P(A \cap B) = P(A|B)P(B)$
2. $P(B|A) = \frac{P(A|B)P(B)}{P(A)}$

3.1.4 INDEPENDENCE AND BAYES RULE

We can take a deck of cards and divide it into four subsets, each set containing all cards of a suit. We would have the set of hearts, the set of spades, the set of clubs and the set of diamonds. The four subsets are disjoint sets and the union of all of them is the deck of cards. This leads us to the concept of a partition of a sample space. A **partition** of a sample space Ω is disjoint events $B_1, B_2, \ldots, B_n$ such that $\Omega = \cup_{i=1}^{n} B_i$:

Let A be any event in Ω. Often, determining $P(A)$ is hard but determining $P(A|B_i)$ is not. The **law of total probability** allows us to compute $P(A)$ based on the partition and is given by Eq. (3.4):

$$P(A) = \sum_{i=1}^{n} P(A|B_i)P(B_i) \tag{3.4}$$

Given the value $P(A|B_i)$, can we use it to compute $P(B_i|A)$? Using the rules of probability combined with the law of total probability we can create what is known as **Bayes' rule**. Bayes' Rule is in Eq. (3.5).

$$P(B_i|A) = \frac{P(A|B_i)P(B_i)}{P(A)} \tag{3.5}$$

If we draw a card from a deck twice, then the second draw is influenced by the first draw. On the other hand, if we flip a coin twice, the outcome of the second flip is not influenced in any way by the first flip. Two events are **independent** if the outcome of the first does not affect the outcome of the second. This implies that if we are considering $P(A|B)$, then the probability of B will not affect the outcome of $P(A|B)$ at all. In other words, $P(A|B) = P(A)$.

Returning to the definition of conditional probability, we have Eq. (3.6):

$$P(A) = P(A|B) = \frac{P(A \cap B)}{P(B)} \tag{3.6}$$

Multiplying both equations by $P(B)$, we see that if two events are independent, then $P(A \cap B) = P(A)P(B)$.

Example 3.1.8. A facility has two independent redundant power sources. It can survive if it loses one, but not both. The first will fail during the week with a probability of 0.1. The second will fail during a week with a probability of 0.25. What is the probability the facility will fail in a week?

Let $P1$ be the event "power source one fails in a week" and $P2$ be the event "power source two fails in a week." We know that $P(P1) = 0.1$ and $P(P2) = 0.25$. Therefore since they are independent, $P(P1 \cap P2) = P(P1)P(P2) = 0.1 \cdot 0.25 = 0.025$.

We know that if two events A and B are independent, then so are A^c and B^c. The pair A^c and B are independent as are A and B^c.

Now suppose we have n independent events, $A_1, \ldots, A_n$. We know that A_i and A_j are independent, so $P(A_i \cap A_j) = P(A_i)P(A_j)$ for all $1 \leq i \leq n$ and $1 \leq j \leq n$. This also holds for any three events, any four events, and even all n events. In other words, since A_i for $1 \leq i \leq n$ are n independent events, then any combination of the events is also independent.

3.2 FROM PARLOR TRICKS TO RANDOM VARIABLES

In probability and statistics, the term **random variable** refers to a variable that takes on values in a numeric sample space to which the rules of probability apply. An event for this variable consists of a numeric set, and probability statements are defined on these numeric sets. So, when X is defined as a random variable, that means it must have two components:

- A set of possible numeric values it can take on.
- A **distribution** over these values, which defines the probability associated with any sub-set $s \in \Omega$ in accordance with the rules of probability that were outlined in Section 3.1. A distribution is a function that quantifies the probability as a function between real numbers.

In essence a random variable is (1) and (2).

Why do we focus on random variables instead of more general variables? The answer is all about the sample space. Random variables have numeric properties that arise from both the sample space and the distribution. Properties like minimum, maximum, median, mean and standard deviation are all properties of numeric random variables.

A variable V that is defined as "the operating system of a desktop computer sampled from the United states" is random, but it is not numeric; it takes on values in an unordered set {Windows, Linux, Unix, ...}. We can describe the probability of any set of events in that space, but it does not make sense for example, to talk about the mean value of V or its standard deviation.

However, an **indicator variable** X that assigns the value 1 to a computer with the Windows OS, and 0 to any other type, is a random variable.

To summarize, a random variable is taking the sample space out of the equation and turning the probability into a function of numbers. Where probability is defined as a function from Ω to $\mathbb{R}$, the random variable is defined as a function where the domain and range is both $\mathbb{R}$.

3.2.1 TYPES OF RANDOM VARIABLES

We have said that a random variable is a function $f : \mathbb{R} \rightarrow \mathbb{R}$. If the range of the function is enumerable, then it is a **discrete** random variable. For example, in Example 3.1.1 we ask the computer for a random integer between 0 and 5. This is a discrete random variable. A discrete random variable cannot take any other values than what it is in the list. The list does not have to be composed of integers, it an be composed of real numbers. The point is that the values of the random variable are restricted to that list.

If the random variable can be any values in an interval in the real numbers, it is a **continuous** random variable. In other words, it is not defined specific values in the interval, it is defined on all values in the interval. If we measure the length of time that a web session lasts, then the outcome of that value is a continuous random variable. The time could be 1 s, or 1.1 s, or even 1.2 s. If our monitor is precise enough, we could even get nanoseconds.

If the random variable is associated with the outcome of an event, we all it an **indicator** random variable. The assigning the value 1 to a computer with the Windows OS and 0 to any other type is an indicator random variable.

3.2.2 PROPERTIES OF RANDOM VARIABLES

It is important to remember that when we say random variable, we actually are talking about a distribution function. It can be confusing otherwise. The underlying process of a random variable is a function that is in essence, a probability. So when we are finding a random value, we are actually finding a probability.

If our random variable is discrete, then the **probability mass function**, also known as the **pmf**, is defined as $f(x) = P(X = x)$ for all $x \in \Omega$. This emphasizes the fact that a random value is a function. To illustrate this, consider the random variable defined by assigning 1 to a computer with the Windows OS and 0 to any other type. The probability mass function of this is $f(1) = P(X = 1) = P(X = \text{Windows})$ and $f(0) = P(X = 0) = P(X = \text{Other})$.

The pmf has two properties. First, $f(x) \geq 0$ for all values of x. Second, if we sum all of the possible values in the pmf, the value must be one.

If the random variable is continuous, then we know that the random variable is a function into the real numbers, which we will denote as $f : \mathbb{R} \to \mathbb{R}$. The **probability distribution function**, or **pdf**, for the random variable is given as $f(x) = P(X = x)$, similar to the pmf.

The pdf has two properties. First, $f(x) \geq 0$ for all values of x. The second is similar to the second property of the pmf. An integral is a special kind of sum, where we are summing the function over every possible value in an interval. In that vein, Eq. (3.7) must hold:

$$\int_{-\infty}^{\infty} f(x)dx = 1 \tag{3.7}$$

This is saying that if we sum our pdf over every possible value we could plug into the function, then that value must be one. To restate this in calculus terms, the area under the curve of the pdf must be one.

The **cumulative distribution function**, or **cdf**, is defined for all possible random variables and is given by $F(x) = P(X \leq x)$. For a discrete random variable, this is translated into a sum over the pmf. Eq. (3.8) is the formula for the cdf for a discrete random variable:

$$F(x) = P(X \leq x) = \sum_{a \leq x} f(a) \tag{3.8}$$

The cumulative distribution function for a continuous function is given in Eq. (3.9):

$$F(x) = P(X \leq a) = \int_{-\infty}^{a} f(x)dx \tag{3.9}$$

This again uses the idea that the integral is in essence a "super sum" over all possible values in the interval. By extending the interval to minus infinity, we are saying sum it over every possible value that x could have up to a.

In both the discrete and random cases, $F(x)$ increases as x does.

Using the cdf, we want to find a value q such that $F(q) = 0.1$. That is, $P(X \leq q) = 0.1$. We know this value exists because the cdf has a domain between 0 and 1, so at some point q, $F(q) = 0.1$. This is called the 10th **quartile**. The cdf is always increasing, so if we consider 0.25, there is a value r such that $P(X \leq r) = 0.25$ and $q \leq r$.

To generalize this notion, we define the pth quartile as the smallest value q_p such that $F(q_p) = P(X \leq q_p) = p$. We are abusing notation quite a bit here, because p is a value between 0 and 1, yet we ignore the decimal and call it the pth quartile. So in our previous example, the 10th quartile is when $p = 0.1$.

The 25th quartile, 50th quartile and 75th quartile are special cases. We denote the 25th quartile by Q1, the 50th quartile by Q2 and the 75th quartile by Q3. The **interquartile range** is given by Q3 − Q1 and denoted *IQR*. *IQR* can be considered as a measure of variability, as it determines the spread of the middle values of the distribution. If it is quite large, then the distribution, at least the middle values, are spread out. If it is small, then the distribution is narrowly clustered around Q2. The value for Q2 is also called the **median** of the distribution.

The **expected value** of a random variable is a summarization of the distribution. It gives the mean, also known as the center, of the distribution of the random variable. If we have $a_1, \ldots, a_n$ discrete random variables and a pmf of $f(x)$, then the expected value of the discrete random variable is given in Eq. (3.10):

$$E(X) = \sum a_i f(a_i) \tag{3.10}$$

It is a weighted average of the random variable and can be thought of as the value we anticipate from the distribution. We anticipate that if $P(1) = 0.9$ and $P(0) = 0.1$, then the value we will most likely see in our distribution is close to 1, as we see from computing the expected value.

Example 3.2.1. Suppose we have a laptop battery that we know will last for 2 h with a probability of 0.7, 3 h with a probability of 0.5 and 4 h with a probability of 0.1. We want to have a general idea of about how long this laptop battery will last.

The answer is the expected value. What do we expect will happen and how long do we expect it to last? In this case, our pmf is given by $f(2) = 0.7, f(3) = 0.5$ and $f(4) = 0.1$. We use Eq. (3.10) to determine that the answer is 3.3. We expect that the battery will last longer than 3 h but less than 4.

We extend this to the continuous case using the pdf of the continuous random variable. Eq. (3.11) is the expected value for the pdf $f(x)$:

$$E(x) = \int_{-\infty}^{\infty} xf(x)dx \tag{3.11}$$

If we sum two random variables, then the expected value of the sum is the sum of the expected values. That is, $E(X + Y) = E(X) + E(Y)$. Similarly, if we have two constants $a, b \in \mathbb{R}$, then $E(a + bX) = a + bE(X)$.

Let us digress for a moment into math. If we have another function $g(x)$ and compose $f(x)$ and $g(x)$ as $f(g(x))$, then we can consider the pdf of $f(g(x))$. Now we are no longer asking $f(x) = P(X \leq x)$ but asking $f(g(x)) = P(X \leq g(x))$. We can also find the expected value of $g(x)$, which we write as $E(g(x))$. It can be shown that $E(g(x)) = \int g(x)f(x)dx$ for a continuous random variable. In the discrete case, it is $E(g(x)) = \sum g(x)f(x)$.

Returning to our discussion of random variables, let us consider the question "how much does this random variable vary from the expected value?" We could have a random variable that is clustered around the mean with only a few values far from it, or we could have one that is distributed equally far on both sides of the mean, but is not close to it at all. The **variance** allows us to quantify the value of "spread about the mean."

The variance is defined as $\text{Var}(X) = E(X - E(X)^2) = E(X^2) - E(X)^2$ and is always greater than 0. If it is small, then the values of the random variable are close to the expected value. If it is large, then the values are more spread out.

The **standard deviation** is defined as the square root of the variance. This is a similar measure to the variance and also determines how spread out the random variables are. The standard deviation is denoted as σ and the variation is also denoted as σ_X^2 as well as $\text{Var}(X)$. In an abuse of notation, we usually leave the X off and denote it as σ^2 or σ if we are considering a single random variable.

As we saw in the expected value, the sum of the variances of two random numbers is the variance of the sum. To be precise, $\sigma_{X+Y}^2 = \sigma_X^2 + \sigma_Y^2$. On the other hand, if we have to constants $a, b \in \mathbb{R}$, then $\sigma_{a+bX}^2 = b^2\sigma_X^2$. It is important to note that the sum property does not necessarily hold in the standard deviation. We can see this in Eq. (3.12):

$$\sigma_{X+Y} = \sqrt{\sigma_{X+Y}^2} = \sqrt{\sigma_X^2 + \sigma_Y^2} \tag{3.12}$$

The square root of the sum of two values is not necessarily equal to the sum of the square root, so the sum property does not hold for standard deviations.

Variances and standard deviations are hard to compute sometimes, as they are defined on $E(X^2)$ which is not always tractable. In math, we often take these problems and recast them in such a way that they become solvable. The same holds true in this case. The values $E(X), E(X^2), E(X^3), \ldots, E(X^n)$ are called **moments**. $E(X)$ is the first moment, $E(X^2)$ is the second moment, and $E(X^n)$ is the nth moment.

The **moment generating function** of a random variable is given by $M(t) = E(e^{tx})$. If X is a discrete random variable with pmf $f(x)$, then the moment generating function of f is given by Eq. (3.13):

$$M(t) = E(e^{tx}) = \sum_{x \in X} e^{tx} f(x) \tag{3.13}$$

In the continuous random case, if we have a pdf of $f(x)$ then the moment generating function is in Eq. (3.14):

$$M(t) = E(e^{tx}) = \int_{-\infty}^{\infty} e^{tx} f(x) dx \tag{3.14}$$

The reason it is called the moment generating function is because we can generate the moments using it. Our first claim is that $M'(0) = E(X)$. This means if we take the first derivative of the moment generating function and evaluate it at 0, we get the expected value of X as a result. If we take the second derivative and evaluate it at 0, we get the second moment as the result. That is, $M''(0) = E(X^2)$. We can then compute the variance as $\sigma^2 = M''(0) - (M'(0))^2$.

We have recast the problem of finding the moments of a function to using the moment generating function, which is generally easier to compute. This allows us to compute the variance and standard deviation, as well as the expectation, even in cases where the integral is difficult to compute.

3.3 THE RANDOM VARIABLE AS A MODEL

So now that we have spent time discussing random variables, let us talk about why we want to use them. Informally, the random variable describes the likelihood for it to assume a particular value. We began with random events and sample spaces, removed sample spaces from the discussion and created our real value functions called random variables. This implies that our random variables can be used to model random events.

If we create a random variable for each possible random event, we will spend a lot of time reinventing the wheel, and also running into cases that we can just not compute. To get around this, we have a collection of known distributions of random variables. These distributions have properties which are either inherent in the distribution or defined over time. If we can show our random event "fits" a random distribution, then we can use it and the properties of it in order to model our random event. Model fitting is where we take our data and the distribution, and determine how close the data in the distribution is to our data.

In this section we will discuss common distributions of random variables and their properties.

3.3.1 BERNOULLI AND GEOMETRIC DISTRIBUTIONS

Suppose we have a server that we are continually pinging. The server can be up or down, but no state in between. For the sake of this discussion, we will assume that the outcomes up and down are independent. In other words, the fact that the server is down one ping does not influence whether or not the server is up on the next. This experiment is very similar to other experiment that we can perform repeatedly with two independent outcomes, such as flipping a coin, a true/false test, or even the question "Is the computer infected with a virus?" We can model these experiments with a discrete random distribution called the **Bernoulli** distribution. The Bernoulli distribution is the simplest random distribution and can be used to build other distributions. The random experiment that creates the Bernoulli distribution is called the **Bernoulli trial**.

The Bernoulli distribution models a random event X that can have two outcomes, which we will call a success and a failure. We define one as success, zero as failure, so $P(X = 1) = p$ and $P(X = 0) = q$. In other words, our pmf is given by $f(1) = p$ and $f(0) = q$.

We know that the sum of the values of the pmf is 1, so $q = 1 - p$. Eq. (3.15) computes the expected value for the distribution:

$$E(X) = \sum a_i f(a_i) = 0\dot{q} + 1\dot{p} = p \tag{3.15}$$

The Variance is also easy to compute, it is $\sigma^2 = pq = p(1 - p)$ and the standard deviation is $\sigma = \sqrt{p(1 - p)}$.

The random experiment that creates the Bernoulli distribution is called the **Bernoulli trial**.

The **geometric** distribution is based on the Bernoulli distribution. It answers the question "what is the probability that our first success will be on the jth trial?," that is, we are looking for $P(W = j)$ where W is an integer representing the trial. The formula for the pmf of the geometric distribution is $P(W = j) = q^{j-1}p$. It is a discrete distribution because we can list the possible results for the function.

3.3.2 BINOMIAL DISTRIBUTION

Similar to the geometric distribution, the **binomial** distribution is based on Bernoulli trials. We assume that each trial is a separate event and the probabilities remain constant regardless of past trials.

Let X be the number of successes in n trials. In the geometric distribution, we were asking the question related to finding the first success, in the binomial distribution we are asking what is the probability of j successes in these n trials. In this case, the pmf is given by $f(j) = P(X = j) = \binom{n}{j} p^j q^{n-1}$. It is a discrete random variable, because we can list the possible values that $f(x)$ can achieve.

The cdf of the binomial distribution is $F(k) = P(X \leq k) = \sum_{i=1}^{n} \binom{n}{i} p^i (1 - p)^{n-i}$.

Clearly, the variable n is a parameter for the binomial distribution and must be chosen at the beginning. The values we obtain in the distribution depend on this parameter. Similarly, the probability p also influences the outcome.

The parameter influences the expected value of the distribution. With a bit of algebra, we can show that $E(X) = np$. Similarly, the variance of the binomial distribution is $\sigma^2 = np(1-p)$ and the standard deviation is $\sigma = \sqrt{np(1-p)}$.

Example 3.3.1. Suppose we have a process that cleans an infected server with probability of success as 0.9 and a probability of failure as 0.1. In other words, $P(1) = 0.9$ and $P(0) = 0.1$. If we have 20 infected servers, how many will be cleaned using this process? The expected value of this distribution is $E(X) = np = 20 \times 0.9 = 18$.

3.3.3 POISSON DISTRIBUTION

The binomial distribution is relatively easy to compute when the number of trials is a reasonable number, but what happens if n is in the millions? Computing $\binom{n}{k}$ when n is very large is a difficult proposition. In math, we often use approximations in these cases, where the approximation gives us an answer that is close to the correct one, if not exact. The **Poisson** distribution is an approximation of the binomial distribution for large values of n.

Choose a fixed constant $\lambda > 0$. Then the Poisson distribution is given by Eq. (3.16). This constant is a parameter of the Poisson distribution, similar to the n and p of the binomial distribution:

$$f(j) = P(X = j) = \frac{\lambda^j}{j!}\mathrm{e}^{-\lambda} \tag{3.16}$$

The Poisson distribution is a discrete distribution because we can list the values associated with it.

Example 3.3.2. Suppose that the number of servers that fall off the network during a 24 h period can be modeled with the Poisson distribution with a parameter of 1.5. What is the probability that three or fewer servers will fall off the network in 24 h?

The answer to this question is $P(X \leq 3)$ where P is the Poisson distribution. Since the Poisson distribution is discrete, then $P(X \leq 3)$ is the sum of $P(X = i)$ for $i = 0, 1, 2, 3$. We demonstrate the computation of the cdf in Eq. (3.17):

$$P(X \leq 3) = \frac{1.5^0}{0!}\mathrm{e}^{-1.5} + \frac{1.5^1}{1!}\mathrm{e}^{-1.5} + \frac{1.5^2}{2!}\mathrm{e}^{-1.5} + \frac{1.5^3}{3!}\mathrm{e}^{-1.5} = 0.9343 \tag{3.17}$$

3.3.4 NORMAL DISTRIBUTION

All of the random distributions discussed so far have been discrete. We turn our attention to the continuous distribution, by considering the **normal** distribution. The mathematician Gauss used it to model observational errors in Astronomy, so it is often referred to as a **Gaussian** distribution.

The distribution is a symmetric distribution, with a central peak that has equal values on each side. The peak is given by the mean of the distribution and the shape of the curve is defined by the variance. This means that the two parameters that define the distribution are the mean and the standard deviation. We denote the mean by μ, so the pdf of the normal distribution with mean μ and standard deviation σ denoted by $N(\mu, \sigma^2)$. The pdf is given in Eq. (3.18):

$$N(\mu, \sigma^2) = f(x) = \frac{1}{\sigma\sqrt{2\pi}} e^{-\frac{1}{2}\left(\frac{x-\mu}{\sigma}\right)^2} \tag{3.18}$$

Since we define the expected value and the standard deviation as the parameters of the normal distribution, it works out so that they are the actual expected value and standard deviation. Fig. 3.1 is a plot of a normal distribution with a mean of 0 and a standard deviation of 1.

This is a popular version of the normal distribution because any $N(\mu, \sigma^2)$ can be transformed into $N(0, 1)$.

There is an interesting property of the normal distribution that applies to any normal distribution, no matter what the parameters are. If we consider the interval $[\mu - \sigma, \mu + \sigma]$ of the distribution, then 68% of the values in the distribution appear in that interval. Extending that interval to $[\mu - 2 \cdot \sigma, \mu + 2 \cdot \sigma]$, then 95% of the values of the distribution appear in that interval. Now taking it one step further, to $[\mu - 3 \cdot \sigma, \mu + 3 \cdot \sigma]$, 99.7% of the values appear in that interval. This property does not hold for other distributions, only for the normal distribution.

If we are using the normal distribution to model errors in TCP transmissions, then 68% of those errors appear in the interval $[\mu - \sigma, \mu + \sigma]$.

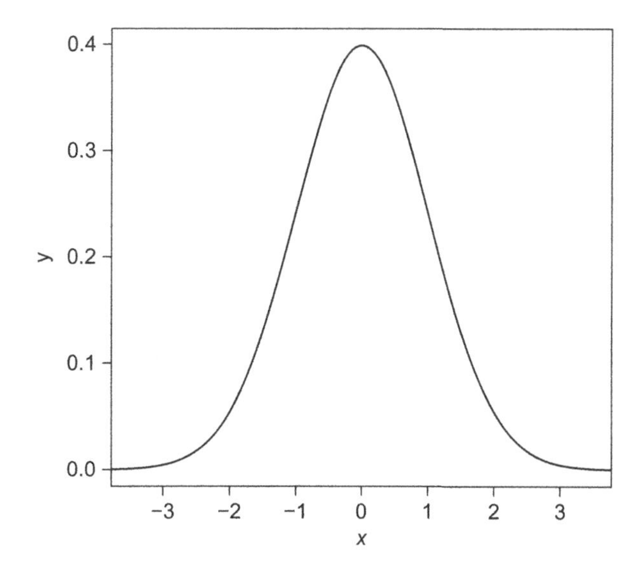

FIG. 3.1

The normal distribution with $\mu = 0$ and $\sigma = 1$.

3.3.5 PARETO DISTRIBUTIONS

An economist named Pareto noticed more than a 100 years ago that the number of people whose incomes were greater than a value x could be modeled by the function $\frac{C}{x^\alpha}$ for constants $C > 0$ and $\alpha > 0$. This is called a **power law** distribution. A power law distribution has the property that large numbers are rare, but smaller numbers are more common. So it is more common for a person to make a small amount of money versus a large amount of money.

The **Pareto** distribution is a continuous power law distribution that is based on the observations that Pareto made. The pdf for it is given by $f(x) = \frac{\alpha}{x^{\alpha+1}}$ and the cdf is given by $F(x) = 1 - \frac{1}{x^\alpha}$.

The expected value of the function is based on the parameter. If $\alpha \leq 1$, then the expected value of the Pareto function is ∞, or infinity. If it is greater than 1, then it is given by $E(X) = \frac{\alpha x}{\alpha - 1}$. Similarly, the variance is based on α as well. If $\alpha \leq 2$, then the standard deviation is ∞ as well. We can interpret that to mean that the values of the distribution are spread very far out from the mean, with little to no clustering. If the value of α is greater than 2, then $\sigma = (\frac{x}{\alpha-1})^2 \frac{\alpha}{\alpha-2}$.

Fig. 3.2 is the general shape of the Pareto distribution. If x is close to 1, then the probability of it occurring is much higher than for a value farther to the right on the graph. If x is large, then the probability is closer and closer to 0.

This is true for most power law distributions. A discrete variant of the power law distribution is called **Zipf's** distribution. The math required to understand the pmf of the distribution is beyond this book, however, and it will not be covered.

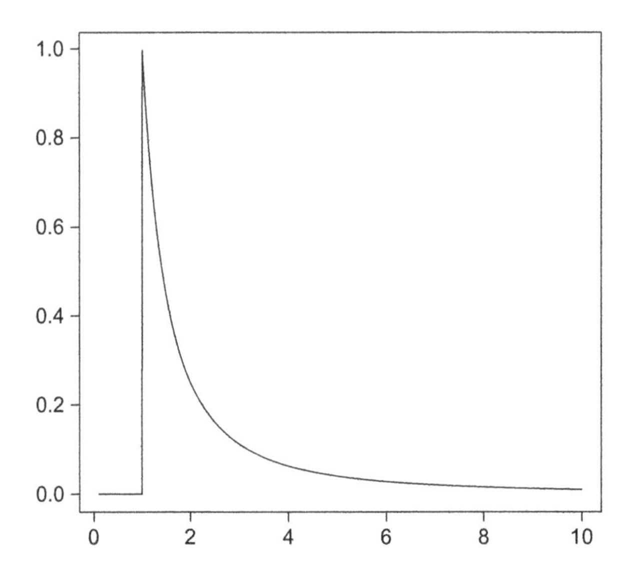

FIG. 3.2

The Pareto distribution.

An example of the power law distribution is the number of IP addresses associated to a computer systems. Often just one IP address is associated with a given system, however, some may have thousands or even more. We could model this distribution using a power law.

3.3.6 UNIFORM DISTRIBUTION

If we wish to find a distribution that gives a random number between two given numbers, then we want the value to be anywhere in between the two endpoints, not focused on the mean, which occurs with the normal distribution or on the extremes, which occurs with a power law distribution.

The **uniform** distribution is a continuous distribution that fits this requirement. It is defined on an interval $[a, b]$ where $a, b \in \mathbb{R}$ are chosen as our limits. We want every value of our pdf on the interval to have an equal value, so the pdf for the distribution is given in Eq. (3.19):

$$f(x) = \begin{cases} 0 & x < a \\ \frac{1}{b-a} & a \leq x \leq b \\ 0 & x > b \end{cases} \tag{3.19}$$

The cdf is then given in Eq. (3.20):

$$F(x) = \begin{cases} 0 & x < a \\ \frac{x-a}{b-a} & a \leq x \leq b \\ 1 & x > b \end{cases} \tag{3.20}$$

This makes sense because as the value of x grows, the cdf $F(x)$ should approach the value 1. In the case of the uniform distribution, that occurs as x nears b.

The expected value and the variance of the uniform distribution are also equally easy to compute. $E(X) = \frac{1}{2}(b - a)$, or the midpoint of the interval. This is logical as the expected value is the mean, or the midpoint and therefore $E(X)$ should be the midpoint. With a bit of algebra, we can compute the variance as $\sigma^2 = \frac{1}{12}(b - a)^2$. If the interval is small, then the variance is small, as there is not much room to spread out. As the interval grows, we expect the variance to grow as well.

3.4 MULTIPLE RANDOM VARIABLES

The random variables discussed in this chapter have been singular, meaning we have only discussed single random variable distributions. Now we can consider the case of two random variable distributions. The univariate (or single value) distributions were a function from $f : \mathbb{R} \to \mathbb{R}$. The multivariate (or multiple value) distributions are defined as $f : \mathbb{R}^n \to \mathbb{R}$.

For simplicity's sake, we will consider the case where $n = 2$. This keeps our notation simple. Let us return to the premise in Example 3.1.5 where we ask a computer for two integers. We will let X be the sample space defined by the sum of the two integers and Y be the sample space defined by the difference of the two integers. The joint distribution we saw before was $P(X = n, Y = m)$. Turning this into a random variable, we are finding the joint random distribution $f(x, y) = P(X = n, Y = m)$. This function is the **joint probability mass function**.

In a similar fashion, we can define the **joint probability density function** for a continuous function. This is a function $f : \mathbb{R}^2 \to \mathbb{R}$ denoted by $f(x, y)$. The function must always be greater than or equal to 0 and Eq. (3.21) must hold:

$$\int_{-\infty}^{\infty} \int_{-\infty}^{\infty} f(x, y) dy dx = 1 \tag{3.21}$$

The **joint cumulative density function** is given by Eq. (3.22):

$$F(a, b) = P(X \leq a, Y \leq b) = \int_{-\infty}^{a} \int_{-\infty}^{b} f(x, y) dy dx \tag{3.22}$$

If we have the density function for the pair, we can derive the density function for each individual variable. These are generally called the **marginal** density. Eq. (3.23) are the formulas for the marginal density:

$$F(x) = \int_{-\infty}^{\infty} f(x, y) dy \tag{3.23}$$

$$G(y) = \int_{-\infty}^{\infty} f(x, y) dx \tag{3.24}$$

3.5 USING PROBABILITY AND RANDOM DISTRIBUTIONS

We defined random variables based on random events in Section 3.2. In short we used the probability of random events to create random variables and defined properties of the random variables. In Section 3.3, we discussed various random variable distributions. These had functions that defined them as well as parameters that were used to define their behavior. It is quite obvious that the normal distribution has a different shape, and therefore a different use, from the Pareto distribution.

Now we are going to consider how we can use the random distributions. Let us suppose we have a set of data that was created by some random experiment. We know nothing about the way the probabilities were computed for our data, just that these numbers are associated with the set of data. For example, suppose we collect the session times when people access a web server or the number of packets that access a given port on a system. In order to analyze our data set, we want to use a random variable distribution that we know something about. If we determine that the distribution is normal, then we know that 68% of the sessions last between $[\mu - \sigma + \mu + \sigma]$. On the other hand, if we can determine that it is a power law distribution, then we know most of the sessions are very short while a few last a long time.

So we have two questions to solve. First, we need to determine which distribution best fits our data. Second, we need to determine the parameters of that distribution that best fits our data. This is the point where statistics and probability meet. We use the random variable distributions in probability to describe the data we want to analyze in statistics.

Example 3.5.1. Suppose that over a period of time, we have pinged a host to determine if it is up or down. After pinging it 1000 times, 95% of the results are up while 5% of the results are down. How can we best model this data set?

We begin with the Bernoulli distribution, as it is designed to model just this situation. We determine that $p = 0.95$ and $q = 0.05$.

Now if we wanted to ask the question, "given 1000 pings, what is the probability that 950 will be a success?" we would use the Binomial distribution to model the problem and determine the solution.

The first question is picking the distribution that best describes our data. As we saw in Example 3.5.1, that can be as easy as examining the data. Unfortunately, that is not often the case. The problem of determining the best distribution is actually intertwined with finding the best parameters for the distribution, then computing the error. The error can be considered as "how badly did we fit?" the data to the distribution. If the error is small, we did a good job. If it is not, we did not.

Often, when fitting a distribution, we must check more than one distribution. Analyzing the data prior to picking a distribution is a good idea and we discuss this extensively in Chapter 4. This analysis is a statistical method, which once again demonstrates how the two fields are intertwined.

There are many methods for analyzing the fit of a distribution and the parameters of the distribution. The fit of the distribution is determined by using various tests, which in some cases can give misleading results. This is where judgment comes in. An experienced statistician might compare plots of data versus the distribution to determine if the two are similar. Similarly, there are "goodness of fit" tests which also determine if a sample of data originated from a specific distribution.

Once we have found the best distribution and parameters for our data, we can now ask questions of the data set. We can also ask questions about future experiments, given that the current distribution and parameters apply, as we did in Example 3.5.1.

The math behind many of the tests is quite extensive, and complex. For this introduction, the goal for the reader is to be aware that they exist, and that they are used by statisticians and probabilists determine if a collection of data best fits a given distribution. The distribution is then used to make predictions, categorize data, or examine paired events within the data.

3.6 CONCLUSION

Probability is an excellent source of parlor tricks, so that if one has a standard deck of cards, we can determine the probability of drawing five cards of the same suit or how many ways can a deck of 52 cards be shuffled. Interestingly, there are more ways to shuffle a deck of 52 cards than there are stars in the known universe.

Aside from parlor tricks, we discussed in this chapter how to move from probability to random variables. The random variables can be either discrete, meaning they have a discrete number of values in the range of the function, or continuous, where the range of the function is all real numbers in an interval. We then discussed random variable distributions and covered several popular distributions.

The reason we do this is that we are continually studying data and it is very useful to find the random distribution that best describes the events that produced the data. In Chapter 4, we cover further how probability and statistics are intertwined through Data Analysis techniques.

CHAPTER

Introduction to data analysis

4

We use statistics to study the collection, analysis and interpretation of data. It is a collection of methods used for data analysis that is used in a wide variety of fields, from politics to game theory to cybersecurity. It is also a field that is rife with formulas.

We focus in this chapter on the concepts behind various key formulas. The point is to understand why we need to use these formulas. It is less important to memorize them. We will also create a basic toolkit for exploratory analysis and simple visualization of data.

The goal of this chapter is to give a starting point to understanding statistics, not to give an in depth introduction to the field. We want to talk about the language of data analysis, emphasizing exploratory data analysis.

4.1 THE LANGUAGE OF DATA ANALYSIS

A collection of data requires analysis to tell us something. Otherwise, all we have is a set of data points that was collected in somehow, but we know nothing about the data itself. For example, if we collect passive DNS data over 1 day, we know that during that day, those domains resolved but this does not tell a story about the domains. Analyzing data requires us to collect it, summarize it, and interpret the results.

Summarizing data can include filtering or categorizing our data. For example, to classify our passive DNS data set into two categories, "new domains" and "old domains."

Interpreting our results can allow us to use the data we collect in various ways. We may be able to use a data set to predict future events, such as using a collection of network traffic over 1 year to predict next year's traffic pattern.

We can also use data to theorize hypotheses and test them. Causal relationships are when two events which are connected by a cause and effect. An example of a causal relationship is the presence of malware on a network. The malware is the cause, the effect can be the actions of the malware, such as data exfiltration. We would see the effect in the network traffic of a system as the data is copied from the

Cybersecurity and Applied Mathematics. http://dx.doi.org/10.1016/B978-0-12-804452-0.00004-X

system. We can also theorize that in the network traffic, a spike is directly caused by a virus and then test this hypothesis with the available data.

Equally important, we can use data analysis methods to inform policy decisions. As much as possible, we should leverage the data we collect, such as network traffic, to assist decision makers in designing relevant policies for cybersecurity.

It is important to recognize that all data has uncertainty. We will never have a perfect data set. That is, there is always imperfect information contained within the data. Returning to our example of passive DNS, we cannot assume that that data is perfectly correct. For example, a badly configured nameserver may return results for domains that do not exist. This data can affect the outcome of analysis, so we must account for some degree of uncertainty in our data. What we want to do is to generalize the stable features as well as analyze the stable trends.

4.1.1 PRODUCING DATA

The process of analysis begins when we decide what group we want to study or learn something about. This group is called the **population**. This word does not necessarily refer to just people, but is used in a broader sense. We want to use it to refer to people, animals, things, domains, malicious software, and more. For example, we could be interested in:

- The opinions of the US population about the security of their online banking.
- How the population of nameservers is affected by the introduction of passive DNS collection.
- The average price of firewall protection for all Windows systems.

In summary, the population is the entire group that is the target of our interest. In most cases, the population is dynamic. For example, the amount of traffic on a network has a past, present and future. We want to be able to say something about the future of the network, based on the past and the present.

The population can be very small, such as "my home network today and in the future." Or it could also be as large as the whole Internet, or United States population, or the set of all authors of malicious software. It could be one thing or many things.

The population can be so large that as much as we want to, there is no way we can study all of it. If we tried to poll absolutely every adult in the United States on their opinion of online banking security, we would never finish. Similarly, if we tried to examine every single domain name in existence. A more practical approach would be to example and collect data only from a sub-group of the population, which we call a **sample**. We call this first step, which involves choosing a sample and collecting data from it, **Producing Data**.

When practical reasons demand that we collect a sub-group of the population rather than the whole population, we should make an effort to choose a sample in such a way that it will represent the population well. For example, if we are examining all domain names, we should not restrict our collection to one particular top-level domain.

In the world of cybersecurity data, we can often gather large amounts of data that is immediately observable, but it is expensive to ask questions of everything. That is, to find "ground truth" data. To find ground truth is to determine the accuracy of the sample. For example, we can record a set of IP addresses that by some collection of rules are beaconing (ie, sending traffic on regular intervals). Verifying that IP address so that it is not a false positive, however, could be more difficult and time consuming and may not be possible.

4.1.2 EXPLORATORY DATA ANALYSIS

Once the data has been collected, we have a long list of answers to questions, or numbers, and in order to make sense of the data, we need to summarize it in a meaningful way. This is the second step in the Data Analysis process is called **Exploratory Data Analysis**.

When we have research questions about our data, we need to be able to translate these into questions that can be answered by our data. For example, this includes quantifying qualitative thoughts into quantitative data summaries. If our hypothesis is most of the population of United States adults is concerned about the security of their online banking, then we can easily quantify that into a statement about our data. In other words, we take a hypothesis about our population and reconstitute it as a statement that can be quantified by the data.

Often this involves creating a visualization of the data. The visualization is dependent on the kind of data that we have and can be used to locate anomalies in the data set, as well as find trends within the data. For example, if we collect network traffic every day for a week and then visualize it, that 1 day where the amount of traffic spikes is an immediate anomaly that should be investigated.

4.1.3 INFERENCE

Obtaining sample results and summarizing them is not the only goal. We also want to study the population, not just the sample, and draw conclusions about the population based on the sample. To do this, we need to look at how the sample we are using may differ from the population as a whole so that we can factor this into our analysis. To examine this difference, we use probability.

In essence, probability is the machinery that allows us to draw the conclusions about the population based on the data collected about the sample. This final of the Data Analysis process is called **Inference**. To summarize, the Data Analysis process starts with a population and the first step is to produce data. This generates data from the population and enables the second step, exploratory data analysis. The third and final step applies probability methods to the data to enable information about the population to be inferred. This is illustrated in Fig. 4.1.

This diagram is a "big picture" and is a good way of thinking about the process. In some situations, we can see exactly and immediately what the data and probability models and inference might be. This could occur with surveys and polls. In other

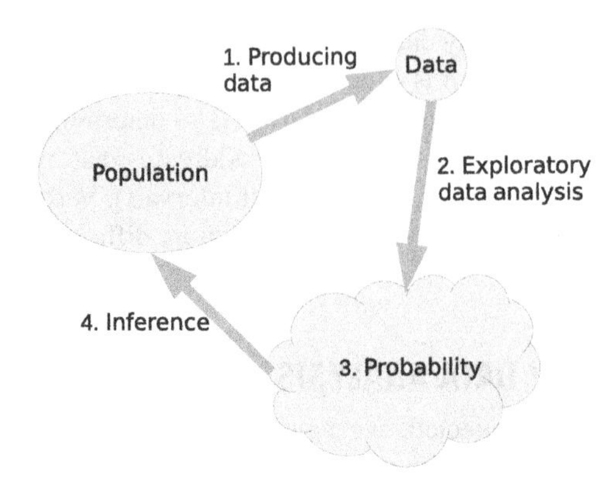

FIG. 4.1

The data analysis process.

cases, it might be more difficult to map our situation to this process, but we can at least start thinking about where and what the difficulties are. It should be used as a guide for the process of statistical analysis.

4.2 UNITS, VARIABLES, AND REPEATED MEASURES

A **unit** is an element of the population that we are studying. If we think of a data set in terms of a database, a unit would be a good candidate for a index. It is also subject to change based on different research questions. Like indices, units can be considered to each have a unique name. For example, a user account in an organization is a unit, a host on a network is a unit, or a domain name is a unit. It depends on the population.

When we measure a property of a unit, it is called a **variable**. The value of a variable can change from unit to unit in the population, hence the name. If one variable is measured per unit, it is called **univariate**. If more than one variable is measured, it is called **multivariate**.

For example, let us consider users and passwords where our unit is the user and the variable is the strength of their password measured weekly for a month. Suppose our results are the results shown in Table 4.1.

We see variation in the measurement across time. Some are weak, some strong, some are medium. They can change from week to week depending on the user. The levels and frequency of the levels are called the **distribution** of the variable.

We can expand this data into a multivariate set. Our units are still the users but our variables are both the strength of the password and the amount of email they send. We summarize this in Table 4.2.

We can summarize each of them separately across our units and we can also study the relationship between the two of the data sets. We study the relationships by

Table 4.1 Password Strength Measured Over Time

	Week 1	Week 2	Week 3	Week 4
Abby	Weak	Weak	Weak	Weak
Bob	Medium	Medium	Strong	Strong
Carol	Weak	Weak	Weak	Weak
Dave	Strong	Strong	Strong	Strong

Table 4.2 Password Strength and Email Traffic Measured Over Time

	Week 1	Week 2	Week 3	Week 4
Abby	Weak 2 msgs	Weak 0 msgs	Weak 10 msgs	Weak 2 msgs
Bob	Medium 143 msgs	Medium 525 msgs	Strong 98 msgs	Strong 102 msgs
Carol	Weak 25 msgs	Weak 32 msgs	Weak 97 msgs	Weak 125 msgs
Dave	Strong 751 msgs	Strong 624 msgs	Strong 590 msgs	Strong 655 msgs

asking questions about the data. Does password strength have any predictive power for the number of emails sent by a user? What about the opposite? Answering these questions are very hard due to the small sample size.

We also have a **repeated measure** in Table 4.2. We measured the same variable multiple times per unit. Two reasons this may be needed are:

- An imprecise instrument leads to measurement error in a fixed value of a variable.
- The variable can change over time when measured in the same unit.

The amount of email a user receives could certainly change over time, as could the strength of their password. This temporal data must be measured repeatedly to see the change over time.

Returning to the question "Is there a relationship between password strength and email volume?" can we say anything useful or interesting about these variables? We have 16 examples of pairs of measurements to examine across four units. Is that enough data to answer our question, or would we be making assumptions about the users based on only a month time?

Suppose we are interested in a policy that would require weekly monitoring and intervention in the event of suspicious activity, where we define suspicious activity as an abnormal amount of email sent. We want our data to inform this policy decision by justifying it in some way by demonstrating that an anomalous amount of email sent is directly correlated to bad behavior.

In some cases, we may not lose any information choosing to summarize this data as though we had a full 16 data points. However, if we start to aggregate across several variables, we may not be able to detect as much on a per-user basis. Some of the variation in pairs of password strength and emails sent is due to repeated measures per user. Detection of an anomalous number of emails sent certainly depends on the user's baseline because what might be anomalous for one user is not for another. We are certainly interested in monitoring this on a per-user basis to find anomalies in emails sent, but not in the collection of pairs that ignores the user and the user's prior history.

In some cases, we may reliably ignore history or between-unit variability, if the size of that variability is small in comparison to the relationship between the variables measured. We need to check that in order to verify the variability and we cannot check it if we do not collect it.

Some common examples of cybersecurity units are IP addresses, account names, domain names, networks, email servers, servers. Common examples of variables, or measurements of this data, are geographic locations, counts of activity, name servers, or data that is associated with timestamps, such as logged in, opened email at a given time, or downloaded suspicious file at a given time.

4.2.1 MEASUREMENT ERROR AND RANDOM VARIATION

Suppose we have a single wood block and we want to measure its width. Our question is "How wide is this wood." Suppose we repeatedly measure it, resulting in a slightly different result each time. These fluctuations come from an imprecise instrument, or from the wood itself. Wood expands and contracts based on external influences. However, if we measure it repeatedly, the error will decrease over time until the average of the measurements is close to the actual width of the wood. Now, suppose we have a collection of wood blocks and a single measurement of each block. Then fluctuations across each measurement are due to the differences in the differences in the widths of the populations of the wood blocks, not from the repeated measures of a single block. The variation in the measures is a value that needs to be examined and explained. A question to consider is "what does the average represent in each case?" In the first example of one block, the average over the measurements is the average width of the single block over repeated measurements. In the second case, the average is the average width over the population of blocks.

Now let us assume we have so many wooden blocks that measuring all of them will take too long. For practical reasons, we choose a sample of the blocks. We need to consider **sampling error**. In its purest form, sampling error is when the population is fixed and the sample is completely representative. In our example of wooden blocks, we begin with a fixed set of blocks that is not changed. So if we choose a random selection of n blocks from the set and measure their widths, then choose another random selection of n blocks from the set, the two averages will be slightly different. These differences are considered the sampling error. In most cases, the sampling error is relatively small and not our most pressing concern. Confidence intervals often represent this kind of variation, that is the uncertainty that arises from observing only a portion of the population. To reiterate, different data sets of the same size may produce different results.

The more pressing issue is whether or not the sample is representative of the population we care about. Our sample should represent the population at large and should not be focused on a single sub-group. It is rare that we have the ability to perform a simple random sample on a population that we care about. It is also rare that we have the ability to sample the entire population that we care about. Consider the population of domain names. Determining all domain names that are registered in the world is a difficult proposition. Not every registrar shares this information freely, so even though it is possible to find out who the registrars are for every top-level domain, it is not necessarily possible to collect all of the registered domain names. We have to make a "best guess" at a sample, hopefully representing all of the top-level domains available.

In the general case, pure sampling error is very rare and is difficult to come by. It shows up mostly in controlled experiments, where we can define the population very well and sample from it easily. In practice, when are we looking at the distributions of variables, we may have all of these sources of variation contributing to the variability we see in our analysis.

Example 4.2.1. We want to build a classifier that would detect malicious websites that are based on "off-by-one" errors. That is, the Levenshtein distance between the website and a known website is one. So we can consider variants of example.com such as exxample.com or eample.com or even esample.com. We begin by considering the active domains that fit this criteria.

Our population is all domains off-by-one from popular sites. The units are the off-by-one domain names. We collect these by first choosing our popular sites. In this case, we use the daily top 500 domains from the Alexa http://www.alexa.com domain ranking. We then use a passive DNS source and collect domains that match our requirements on a daily basis for 85 days. On each day, all unique new domains are compared against the top 500 sites and we then collect any new off-by-one domains. At the end of the data collection, we have 5780 domains.

If we begin with a population, we need to consider how to use it to consider units and measurements. On the other hand, if we begin with a research question before collecting the data, then we can think about the population of interest before we think about how to collect the data. It is generally best to start with considering the research question and considering how to collect the data than to be handed data and have to figure out "what questions can I ask?"

Now let us reconsider the data we have collected. We have a sample size of 5780 domains. Is this representative of my population? We collected from a passive DNS source, is that source stable? Does it change quantitatively over time? We might compare our day to day samples to see if there is a large variance in size to ensure that the data is stable.

We did not start with a population of off-by-one domains and consider a sample of that, we began with a sample collected from the data we had available, also known as a **convenience sample**. So determining a sample error from this data set is not easy as we have nothing to compare it to. The assumptions we are making about this data is that this 85 days of data collection is representative of what we will see in general over time.

We might be able to compare summary statistics of our data against other data sets to see if they are similar or dissimilar, but then again, we may not be able to. We also may not want to. If we are building a method that will use the passive DNS data source, then we probably do not care how it compares to other data sources, we only care about the source we are using. In short, a precise method may not exist for our data and it comes down to our judgment.

4.3 DISTRIBUTIONS OF DATA

In Section 4.2 we said that the levels and their frequencies are called the distribution of the variables. We were discussing the variables found in Table 4.1. To be precise, a distribution is a set of outcomes and an associated frequency for each outcome. They are formal descriptions of variables and summarize the features of the variable in the data. They also describe potential outcomes of a variable on a unit randomly drawn from the population.

There are four basic types of distributions and each relate to the kinds of variables we are examining. If our data has two or more categories associated with it but no ordering, then it is **nominal**. For example, if we are considering network traffic, the protocol associated with the traffic is a nominal variable. Another is the **ordinal** variable, where the data has categories associated with it but the categories can be ordered. In Table 4.1 we considered the password strength of a user as our variable. We can order these variables from "weak" to "strong," so it is ordinal. Both of these distributions are called **categorical** distributions because they use categories.

In categorical data, the **frequency** of a category is the number of variables in that category. Another property of categorical data is the **mode**. That is the category with the highest frequency. Since a frequency is a count, there could be more than one category that has the maximum frequency.

Quantitative variables have numbers we can measure objectively associated with them. There are two variants of quantitative variables, **discrete** and **continuous**. A discrete value is one that cannot be made more precise, for example, the length of a domain name. This is an integer value and can only be an integer. If a variable can take any value between two specified values, it is a continuous variable. Returning to our example of wooden blocks, the width of a wooden block is a continuous variable.

Quantitative distributions have a center and a shape to them. The center can be quantified by one of three ways, the **mean** or average, the **median** which is the middle of the distribution, and the **mode** or the most popular value. For the median, by middle we mean we can order the data and start counting it. If there are 100 variables,

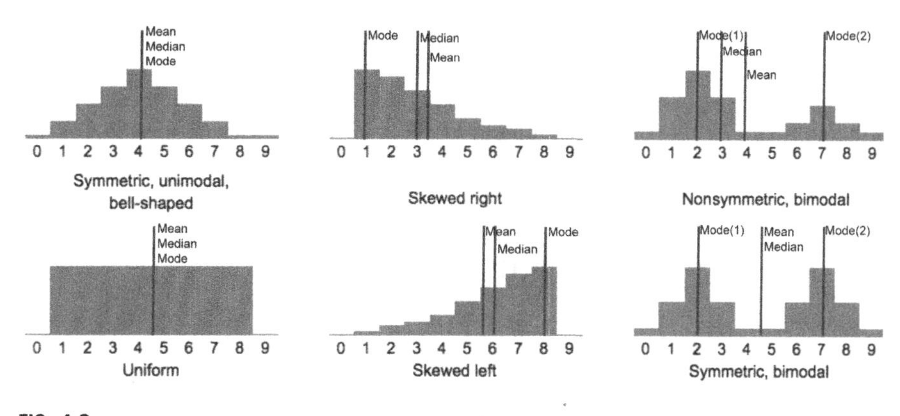

FIG. 4.2

Center and shape of distributions.

then the median is the 50th. The shape can be quantified in multiple ways as well. Fig. 4.2 gives examples of the shape of data combined with the center.

A distribution is **symmetric** if the mean and median occur at the same value. In Fig. 4.2, the top left distribution is a classic symmetric distribution. The distribution is **uniform** if there is no variation in the values, as in the bottom left distribution. Distributions can be skewed right or left, as displayed in the top center and bottom center examples in Fig. 4.2.

A distribution is **unimodal** if it has one mode. Since a mode is the most popular value, we can actually have multiple values as the mode. The symmetric, bimodal distribution on the bottom right of Fig. 4.2 is an example of a distribution with more than one mode.

Determining the shape of the distribution is an important facet to understanding the distribution.

Example 4.3.1. Returning to our off-by-one example, recall that the population is the set of domains that are "off-by-one" character from popular domain names. The sample is the domain names we collect from a passive DNS data source that are off-by-one from a popular domain names.

We can now consider variables associated with these data points. For each domain name, we can determine the action of insertion, deletion or substitution that created it from a popular domain. For example, for exxample.com we know that an insertion of the letter "x" was used to create the new domain. We can then count these to determine the frequency of each category.

We can also consider the rank of the domain that the domain was created from. So if the original domain was ranked n in the Alexa domain rankings, we associate that value to the domain. This is discrete quantitative data.

The rank of a domain in Alexa domain rankings can change from day to day, so we can average its ranking over the time period. This is now a continuous data point.

We do not have repeated measures, since only unique new domains per day were observed. However, we do have repeated measures when we consider popular domains as our population and the top 500 domains as our sample. The new variable is the number of new off-by-one domains, which is now a repeated measure of the count per day.

4.4 VISUALIZING DISTRIBUTIONS

Visualizing the distribution is an important step towards understanding the shape of the distribution. There are many methods to visualize them, this section will cover common methods. Visualizing data is an important part of exploratory data analysis as it can allow us to find anomalies or pattern in the data before moving to probability and inference.

4.4.1 BAR PLOT

A **bar plot** is used to visualize categorical data. We first determine the frequency of the category. A line could be used to display this on the *xy* axis, but to make it clearer, we use a box. We use the data from Example 4.2.1 and consider the number of insertions, deletions and substitutions required to create the new domains. Fig. 4.3 is the result.

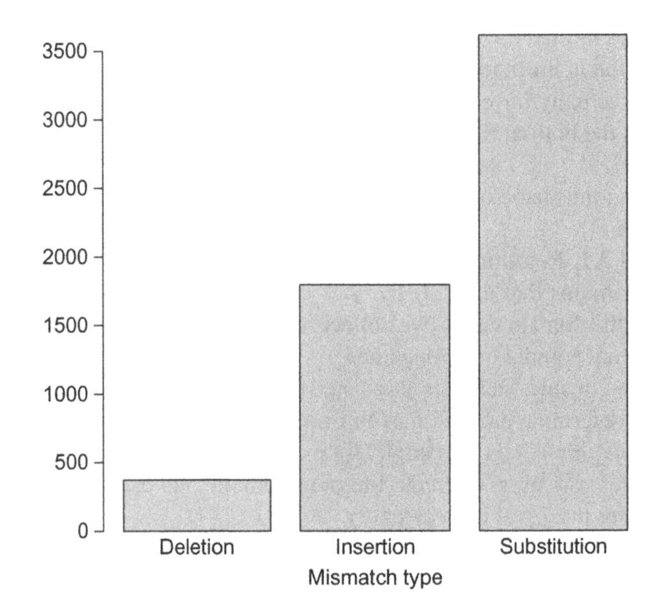

FIG. 4.3

Mismatch type.

As seen in the bar plot, the most common domain name in the off-by-one data set is a name where a character in the original domain name has been substituted by another character. The least common is a domain name in which there has been a deletion of a character from the original domain name. Note that the width of each box in the plot is the same.

We can also visualize this with the **pie chart**. We have the total number of variables, in the case of the off-by-one analysis, we have 5780 variables. We visualize it by letting the angle in a circle represent the total amount of the variable. For example, the angle representing the total number of deletions would be less than 10% of the entire circle. The problem with using a pie chart, though, is if the variables are close in magnitude, it can be difficult to see the differences between them. Bar charts are generally clearer in these cases.

4.4.2 HISTOGRAM

A **histogram** is similar to the box plot, but where the box plot was used for categorical data, the histogram is used for quantitative data. Quantitative data has a range of values, and we divide the range into equal sized bins. Then we count the number of elements in each bin. Recall that this is also known as the frequency. A box is drawn on the plane where the width of the box is the size of the bin and the height is the frequency. Histograms can be drawn either vertically or horizontally.

In a bar plot, the width of the box has no inherent meaning. However, in the histogram, it denotes the width of the bin. If we make the bin narrower, then the amount of elements in the bin could decrease. If we increase it, we could remove all meaning from the plot. Fig. 4.4 is an example of a histogram from Example 4.2.1 that considers the width of the second level domains found in the off-by-one analysis.

If the width of the histogram is too narrow, then it can show too much data and not enough pattern. If it is too wide, it can overly summarize the data.

4.4.3 BOX PLOTS

In a histogram, we summarized our distribution by dividing the values into bins and only displaying the frequencies of the bins. Another way to summarize a distribution is a **box plot**. It is a short summary that covers the range of the data. Recall that the median is half way into the set. The value that is 25% into the set, also known as the first quartile (Q1), and the value that is 75% into the set, also known as the third quartile (Q3). The second quartile (Q2) is the median.

Our plot begins with the range of the values in the distribution. We plot the first quartile and the third quartile, drawing a box where each end is one of the quartiles. We then draw a line through the box denoting where the median is.

Fig. 4.5 is an example of creating a box plot created from a distribution. At the bottom where the arrow is, we see the minimum and maximum value as well as the first and third quartile. The median is also displayed.

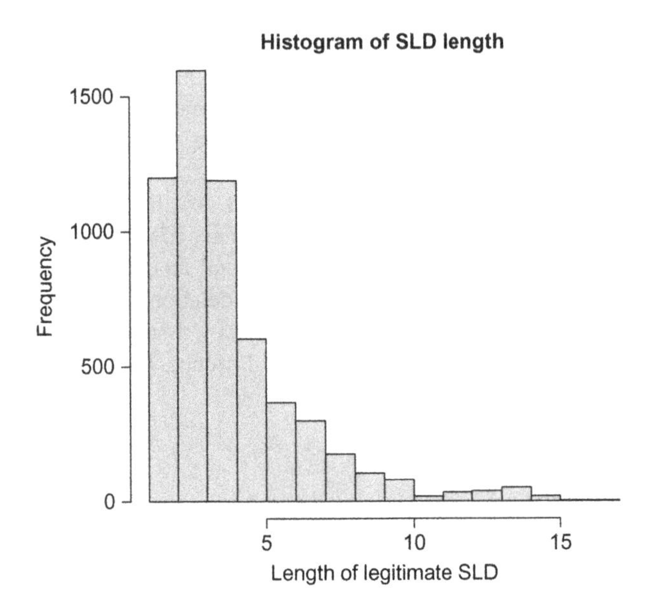

FIG. 4.4

Histogram of second level domain width.

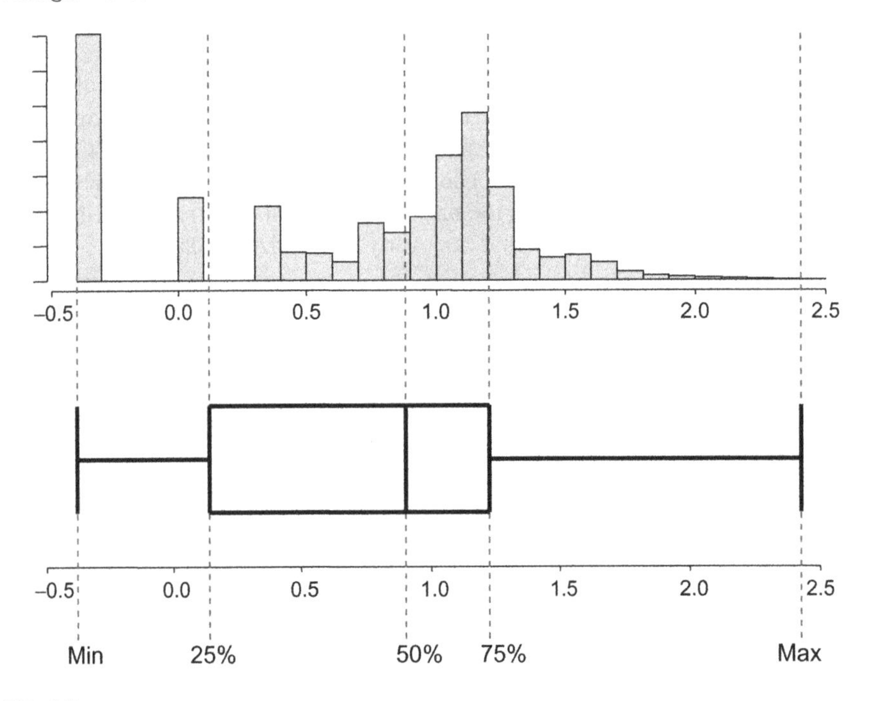

FIG. 4.5

Creating a box plot.

This plot summarizes the data into the range of possible values and the quartiles. A box plot is a way of visualizing the spread of the data from the minimum value to the maximum value and determining where values are grouped in between.

4.4.4 DENSITY PLOT

A **density plot** is a smoothed histogram for continuous variables. There are advantages and disadvantages to the density plot. In a histogram, if the data is fine grained, then identifying a peak can be hard if it always straddles two bins. Another idea is that for continuous measurements, we should have some idea of smoothness or continuity. Imagine, for example, accumulating more and more data up to infinite amounts of data while continually shrinking the bin size of the histogram more and more. We can imagine the histogram converging to some kind of smooth function. In fact, the curves we display for normal distributions as seen in Chapter 3 are exactly these kind of curves. They describe the features of a quantitative variable in an infinite population.

Fig. 4.6 is a histogram of the average rank of the second level domains found in Example 4.2.1 We define the rank of the domain as the rank in the Alexa http://www.alexa.com domain ranking. We can then average these ranks, creating a continuous distribution. Now, we can make the bins smaller and smaller until they disappear, which gives us the density plot in Fig. 4.7.

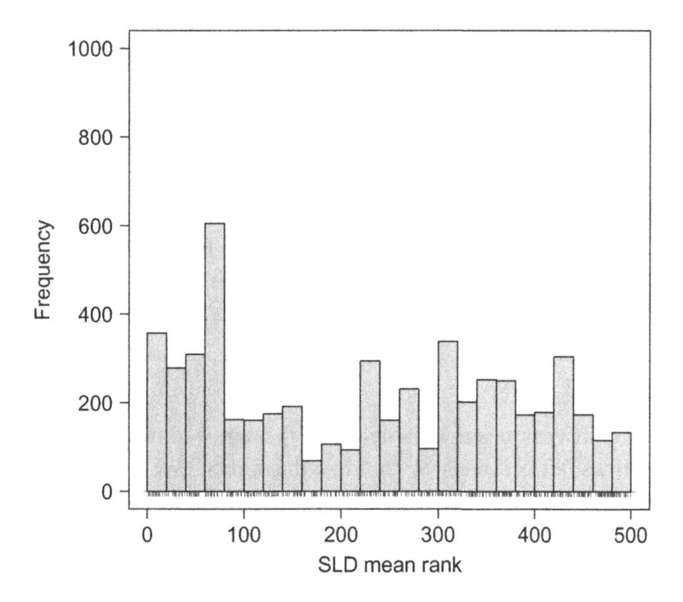

FIG. 4.6

Second level domain mean rank by histogram.

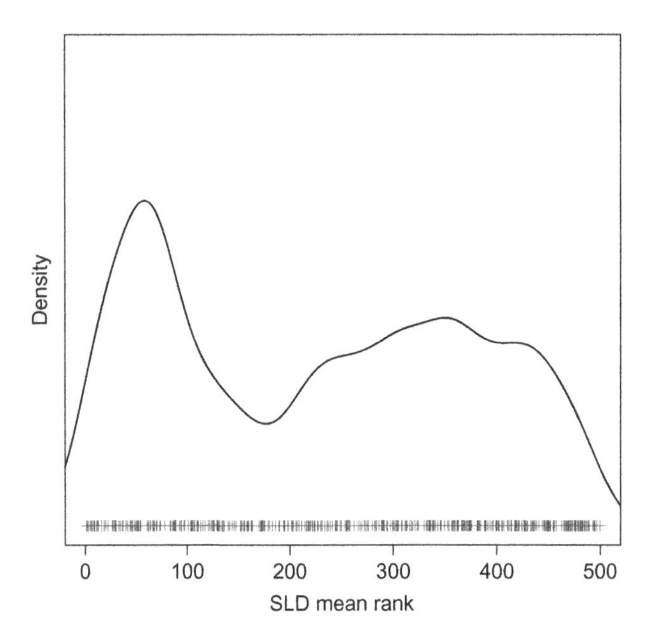

FIG. 4.7

Second level domain mean rank by density plot.

Both figures show essentially the same curve, in Fig. 4.6 it is by box while it is smoother in Fig. 4.7. In this case, we are looking at an estimation of frequency, known as the density. It is also related to the probability density.

A smooth density plot can make it easier to compare the distribution of the data to theoretical density curves, such as the normal distribution. It also uses a kind of a sliding window across the data, so it does not have the difficulty of having to choose break points for bins like the histogram does.

Both plots are exploratory data techniques and both can be helpful. It is a judgment call as to which is best for the data in question.

4.5 DATA OUTLIERS

An **outlier** in a distribution is a variable that is far removed from the set of variables. In essence, it sticks out because it is "not like the others." We can find the outliers by examining data plots or by using methods to examine the data.

We begin by examining how spread out the variables in the distribution are. This is quantified by the **standard deviation**. The standard deviation is a measure of how closely data points are clustered around the mean of the data and is usually denoted by σ. We look at the data points that are further from the mean than three times the standard deviation. We begin with the mean μ and compute $\mu + 3\sigma$. Any values

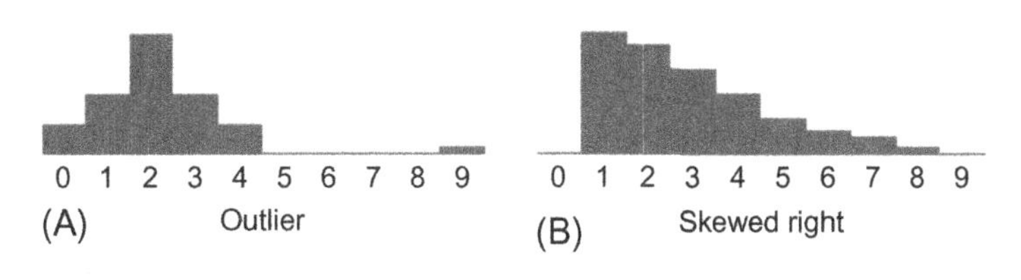

FIG. 4.8

Outliers versus skewed distribution. (A) Outliers. (B) Skewed distribution.

greater than $\mu + 3\sigma$ are considered outliers. Returning to the mean, compute $\mu - 3\sigma$. Outliers are then any values less than $\mu - 3\sigma$.

Another method to find outlier uses the quartile range found in the box plot. After computing Q1 and Q3, that is, the first quartile and the third quartile, we compute the interquartile range, or Q3 − Q1. Now, we can compute Q1 − 1.5(Q3 − Q1) and Q3 + 1.5(Q3 − Q1). Any points outside of these two values, that is, less than Q1 − 1.5(Q3 − Q1) or greater than Q3 + 1.5(Q3 − Q1), are outliers.

It is important to be sure that the points we find are outliers. A skewed distribution may look like outliers, but are actual tails. A **tail** of a distribution is the portion of the distribution that is far from the central part of the distribution, but are not outliers.

Fig. 4.8A shows an outlier to the far right. In Fig. 4.8B we see a value at the same point, but the distribution clearly has a tail.

Outliers can imply experimental error, that is, there was a problem in computing the variable on a data point. These variables are often removed from the data set. It could also be because of a variability within the data set. If our variable is the amount of network traffic on a system every day, then an outlier would indicate that on a given day, that system either recorded an extreme amount of traffic or an extreme lack of traffic. The lack of traffic could be caused by an outage, while the extreme amount of traffic could be caused by a misconfiguration on the system. This example illustrates that not every outlier is a problem with the measurement or necessarily a "bad thing." It does show that outliers should be analyzed.

4.6 LOG TRANSFORMATION

Data transformation is the process of taking a mathematical function and applying it to the data. In this section we discuss a common transformation known as the **log transformation**. Each variable x is replaced with $\log(x)$, where the base of the log is left up to the analyst. It is considered common to use base 10, base 2 and the natural log ln.

This process is useful for compressing the y-axis when plotting histograms. For example, if we have a very large range of data, then smaller values can get overwhelmed by the larger values. Taking the log of each variable enables the visualization to be clearer. An example of this is the number of ports on a system. There are 65,535 ports available on a system, if we are attempting to visualize traffic to all of them, then this visualization could hide values on the lower range of ports while attempting to display higher ports.

Log transformation also de-emphasizes outliers and allows us to potentially obtain a bell-shaped distribution. The idea is that taking the log of the data can restore symmetry to the data.

A log transformation is not always essential to analyzing the data. It can depend on the statistical analysis we are performing. If we are looking for outliers in our data, then a process that de-emphasizes them is not useful. If we use the transformation correctly, then we can illuminate structures in the data we may not have seen. However, we have to be careful on the data that we apply it to. If the distance between each variable is important, then taking the log of the variable skews the distance. Always carefully consider the log transformation and why it is being used before applying it.

4.7 PARAMETRIC FAMILIES

A **parametric family** is a shorthand method for describing common shapes, whether they are discrete or continuous. It can represent a scientific law of how the population should behave and allow us to extrapolate outside observed range. It also allows us to check for outliers as anomalies. The general goal is to find a well-known shape that our data mostly fits, rather than re-inventing the wheel each time we analyze data.

Estimating the shape of the distribution is complex and expensive. The more features that we estimate, the more data we need. Perhaps the distribution fits well within a common family, which would require less work. The difficult part is determine the parametric family with the best fit and finding the parameters within that family.

We call them parametric families because the behavior of members of the family are governed by parameters. The shape of the families are constrained by parameters which can determine variations in the shape. Common families include the Pareto, Gamma, Exponential, Poisson and Normal. Fig. 4.9 is the normal distribution and variations based on the parameters. Each curve is slightly different, yet all of them are within the same family.

When we determine the parametric family that best fits our data, we have created a **model** of the data. We have found a mathematical method that best explains our data and we can use that model to draw conclusions about our population based on the sample. It is the machinery that allows the inference to work.

The aim is to find the parametric family with the best fit to our data. That would give us the family that has the best fit to our data, without leaving important data points out or describing random noise more than the actual relationships within the data. There are a multitude of tools for determining parametric families, we will not

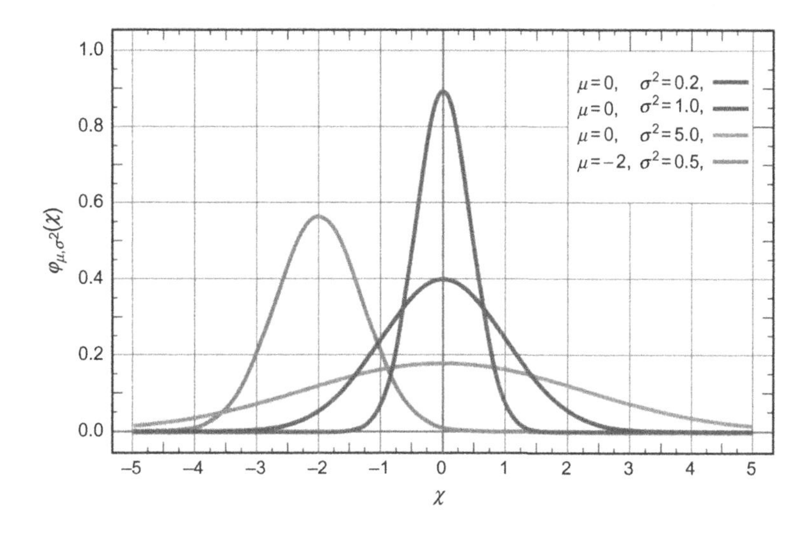

FIG. 4.9

The normal distribution.

cover those. The point, however, is that the probability tool is the engine that drives inference.

4.8 BIVARIATE ANALYSIS

In this chapter we have mostly concerned ourselves with univariate analysis. That is, each data point has one variable associated with it. We now turn ourselves to multivariate analysis, in particular, bivariate analysis. We are considered with two variables associated with each data point.

A **response** variable measures the outcome in a study and the **exploratory** variable is one that influences the response variable. This means that the exploratory variable can cause changes in the response variable. We can use visualization methods to examine relationships between the two variables.

4.8.1 VISUALIZING BIPARTITE VARIABLES

The visualization of the two variables depends on the types, categorical or quantitative, of each variable. If the explanatory variable is a quantitative variable and the response variable is categorical, then we can use side-by-side box plots to visualize their relationships. Similarly, if the explanatory variable is categorical and the response is quantitative, then we use side-by-side box plots as well.

In Fig. 4.10 we have side-by-side box plots to examine the categorical data of deletion, insertion, or substitution from Example 4.2.1. This is the explanatory

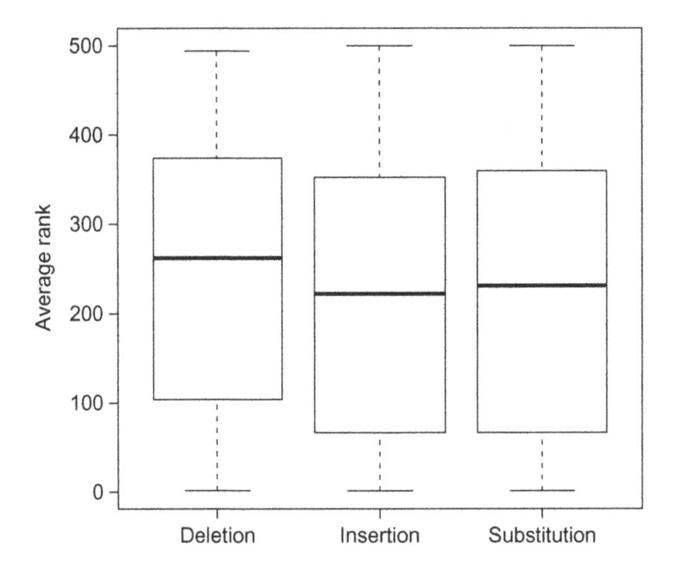

FIG. 4.10

Side-by-side box plots.

variable. The response variable is the average rank in Alexa http://www.alexa.com of the values found in the categorical data.

If the explanatory and response variables are both quantitative, then we use a **scatterplot**. A scatterplot is where we consider the values on the x-axis the explanatory variable and the values on the y-axis the response variable. Each pair (explanatory,response) is plotted on the xy plane. We can also use the scatterplot as a way of categorizing the pairs, as seen in Fig. 4.11. We are comparing the duration of the connection to a webserver with the amount of data that is downloaded.

In the lower left corner we see that there are ephemeral connections, that is, clients that connect to the webserver and download very little. Next to that are automated communicators. These connections last longer than the ephemeral connections but download the same amount. In the middle of the plot are four common downloads. It is easy to combine those if we know the size of files that are commonly downloaded. On the bottom right side we see a very low outlier that falls off the chart. It lasts for a very long time, but downloads very little.

If both the explanatory and response variables are categorical, then we use a **contingency table** to visualize the two variables. A contingency table allows us to examine the relationships between the two categories. A simple example of a contingency table is in Table 4.3. We compare the deletion, insertion, and substitution from Example 4.2.1 and whether not the result of that is not the dictionary, if it is a short word, or a typical word. In this case, the explanatory variable is the deletion, insertion or substitution, and the resulting location of the word is the response variable.

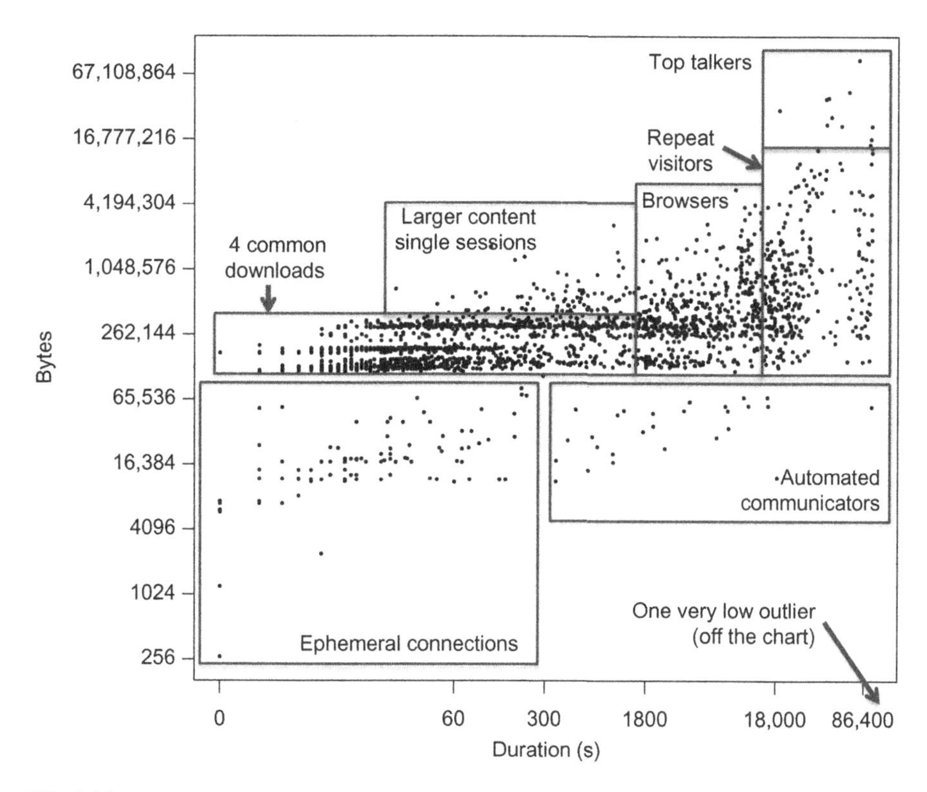

FIG. 4.11

A scatterplot with categorization.

Table 4.3 Contingency Table

	Not in Dictionary	Short Word	Typical Word
Deletion	179	183	10
Insertion	1360	418	17
Substitution	1932	1627	54

4.8.2 CORRELATION

The **correlation** between two variables, in mathematical terms, is the strength of the mathematical relationship between two sets of data. In this chapter, we consider the mathematical relationship to be linear. The correlation boils the relationship between the two sets down to a single number that we can then interpret.

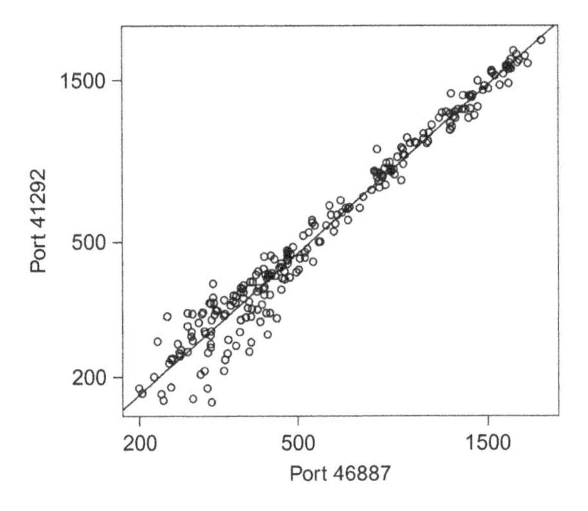

FIG. 4.12

Correlation in a scatterplot.

Visually speaking, if two variables are linearly related, then there is a strong line that runs through the data visualization, in particularly the scatterplot. Fig. 4.12 displays this correlation between ports.

An often quoted maxim in statistics is "Correlation does not mean causation." So if we find correlation between two sets, we cannot assume that this means that the second variable was caused by the first. On the other hand, it is suggestive and implies that the data should be examined further.

4.9 TIME SERIES

In Example 4.2.1 we collected a new set of data every day. We could record those in the order in which we received as successive measurements, keeping track of the time at which each new observation was recorded, as illustrated in Fig. 4.13. This is called a **time series**. Precisely speaking, a time series is a set of quantitative observations that we arrange in order by time.

Time series are generally used for two purposes. On the one hand, we want to understand the underlying process that created the data. In our example, that is the collection from the passive DNS data feed. Our illustration has spikes in the data, we want to know if that is a facet of the collection process or if that is a naturally occurring event. After examining the data, some of the spikes are caused by repeated typos from the same domain. In fact, in the case of one date in particular, there was a large influx of punycode domains that match one domain in particular. However, that does not explain all of the spikes. It might be due to a collection issue on our part, it might be due to a passive DNS data collection error. We do not know.

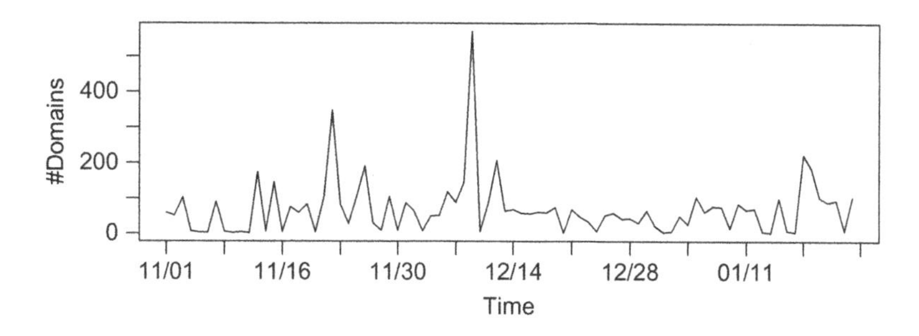

FIG. 4.13

A time series of off by one domains.

Another use for time series analysis is for forecasting or monitoring. We want to use the past performance in order to predict future performance. For example, we can model network traffic as a time series by collecting network traffic periodically. We want to be able to predict how much network capacity we will need in the future based on this data collection.

There are interesting patterns that appear in this data. If we are monitoring an office, then we expect there to be more traffic on Monday through Friday and less traffic on Saturday and Sunday. These repeated patterns can obscure the actual trend in the time series. There are methods to smooth these patterns and find the underlying trends. One in particular is called the Autoregressive Integrated Moving Average, or ARIMA.

4.10 CLASSIFICATION

As we saw in Section 4.5, outliers can be found in data sets. We can examine the outliers and determine why they exist and what they mean. These outliers are anomalies that do not quite fit in with the data set and can tell us about the data. They can also signify problems in the data collection or even problems in our model. The point is that we have to analyze them to determine why they exist.

This brings us to the idea of **classifiers**. This is an idea where we take the data we have and use our model to classify it. It would be nice if there was a model that allowed us to sort network traffic into two buckets, one where one is labeled "good behavior" and the other is "bad behavior." Unfortunately, it is not quite that easy. Some good behavior can look like bad behavior, such as when a user is copying a large file to a customer. This can appear to be data exfiltration. On the other hand, bad behavior can sometimes look like good behavior. DNS has been used for command and control of botnets, so a computer making a DNS query is a common event, not bad behavior. So if we considered the first event as a "bad behavior," we have found a **false positive** once it is examined. It is something that would be classified as bad behavior, but is actually good. The second event is a **false negative**. We have classified it as good behavior wrongly, while it should be bad behavior.

Qualitative variables can also be used to set thresholds in network security analysis. This is a big business, as finding the right threshold that does not overwhelm the analyst with false positives is a very difficult proposition. The threshold should also not obscure false negatives. Finding this value is a common problem in network security monitoring and we can use features of the distribution to aid us in determining the best threshold.

4.11 GENERATING HYPOTHESES

Before we begin by producing data, we need to consider the questions we want to ask. We need to be able to take this question and formulate it in such a way that it is a question about the data, not a general question. This should be the very first step in data analysis. If we do not know our goal, we are certainly not going to be able to get there.

This can then lead us to several questions that should be answered using the data. What summary method do we use with the data? Do we consider the median, the mean, the quartiles? Do we consider frequency counts, or use the histogram method to bin our quantitative data?

The next question is what data should we actually collect to answer our question. We do want a set of "ground truth" data for training, but as we have seen, that is not always possible in cybersecurity data. What structure for the data best handles repeated measures or missing data?

A good investigator always uses exploratory data analysis when beginning the analytical process. The formal tests that are applied when choosing the correct parametric family rely on assumptions that the exploratory data analysis can help verify. Also, standard tools are generally built for the most general problem in mind, not for the specific problem at hand. The more that is known about the current data, the easier it is to customize the data tool and analyze the problem.

So we have produced data, we have used exploratory data analysis, and now we are ready to choose a parametric model, that is, apply probability. We should be careful when we choose a model that we are not modeling the noise or random error in the data set. We would be including bias or sampling error in our trend if that were the case. This is called **overfitting**. On the other hand, we could leave information out of the model, this is called **underfitting**. Both situations modify the results and should be carefully avoided.

4.12 CONCLUSION

Returning to Fig. 4.1, it is important to note that probability and inference only work when the assumptions in producing data and exploratory data analysis are valid. The mathematically inclined might think that the producing data and exploratory data analysis are trivial with the most work for statisticians and data scientists in the probability and inference steps. However, understanding data collection, exploratory analysis and checking assumptions that go into the mechanics of inference is essential to avoid drawing incorrect conclusions. Data is about knowledge, and in many cases, a good picture will illustrate as much as a good statistical test, if not more.

CHAPTER

Graph theory

5

5.1 AN INTRODUCTION TO GRAPH THEORY

A **graph** in mathematics consists of a set of **vertices** and a pairing created with distinct vertices. This pairing creates an **edge**. In visualizing the graph, the vertices are points while the edges are lines connecting two points. The graph is generally written in the form $G = (V, E)$ where V represents the set of vertices and E represents the set of edges. If we let v_1 and v_2 represent vertices, then an edge is written as the pair (v_1, v_2). Then we say that v_1 and v_2 are **connected** by that edge.

Practically speaking, a graph is a way of modeling relationships where vertices can be considered a collection of entities. Entities can be people, systems, routers, DNS names, IP addresses, or malware samples. The edges are then the relationships between two entities. Suppose we have a collection of malicious domains and the IP addresses to which they resolve. Then a domain name has a relationship to an IP address if the domain name resolves to that IP address. In this case, we have an edge between the two, and thus creating a graph that models the domain names and IP addresses of malware. Examining this graph can tell us more about the malware network. Does it use the same IP addresses over and over? Are they scattered, unrelated? Does one IP address serve most of the domains or is it possible that there is one domain that uses a multitude of IP addresses? Analyzing these graphs enables us to answer these questions and more about malicious domains and their IP addresses. The point of using a graph is that we do not need to know much about the malicious software or the domains. We only need to consider the properties of a graph.

We could also draw the graph, as we do for examples in this chapter. This becomes increasingly uninformative once our graph gets large. For a graph with 10 vertices and 20 edges, drawing it out lets us see the important vertices in the graph and potentially the interesting formations in it. However, this becomes increasingly uninformative and complex once graph (ie, the number of vertices) gets large. Drawing a graph with a million edges is nearly impossible by hand. Using math and graph theory allows us to skip the drawing process and summarize information modeled by the graph. Also, we do not need to know what our graph is modeling in

Cybersecurity and Applied Mathematics. http://dx.doi.org/10.1016/B978-0-12-804452-0.00005-1

order to find properties of the graph, we just need the graph. This chapter will cover graphs, their properties, and modeling data with them.

5.2 VARIETIES OF GRAPHS

The types of graphs we will discuss in this section are defined by the data that is being modeled. In other words, we consider the data and the relationships we are examining, and we create a graph that is based on that data and relationships. The varieties of graphs created are selected to highlight various properties found in the graph.

5.2.1 UNDIRECTED GRAPH

Undirected graphs are graphs where the relationship between two vertices is always mutual. That is, if a and b are vertices connected by an edge in an undirected graph, then a is related to b and b is related to a. Undirected graphs are also called **simple** graphs.

A **social network** is a collection of entities, usually people or organizations, that have relationships between them. For example, a high school class is a social network. People in high school have friends, in other words, relationships. A graph can be used to model this situation, as we have entities and relationships. Then the friendship between two people is considered to be a mutual relationship. If you are a friend of Joe in high school, then Joe is also a friend of yours. The peering relationship within BGP is often modeled as an undirected graph. The relationship is not necessarily bidirectional, in that one Autonomous System announces to a second Autonomous System that the first one has the route to the given CIDR block, but it is a mutually beneficial relationship and is often modeled as such.

5.2.2 DIRECTED GRAPH

A **directed** graph is a graph where the relationship between two vertices is a one way relationship. It is the counterpart to the undirected graph. In this case, if a and b are vertices in a directed graph where a is connected to b, then a has a relationship to b but b does not necessarily have a relationship with a. For example, if we are modeling network flow, it has a source IP and a destination IP. We could use a directed graph where the relationship starts at the source IP and ends at the destination IP.

When we visualize this graph, the edges are connected with an arrow. This arrow demonstrates that the origin of the arrow has the relationship with the destination, but the reverse is not necessarily true.

In the remainder of this chapter, when we say graph, we mean undirected graph. We will specify directed graph if it is needed to model the data.

5.2.3 MULTIGRAPH

In a standard undirected graph, there is only one edge between two vertices. Similarly for a directed graph, one vertex can be connected to a second vertex and vice versa, but each relationship occurs only once. A **multigraph** is a graph, either directed or undirected, where the relationship between two vertices is repeated. This has applications when we are modeling network flow records. If there are multiple connections between two IP addresses in the set of flows under examination, a multigraph would be useful. The multigraph would then have an edge between two IP addresses for each connection.

In general, the edges in a graph are between distinct vertices. A multigraph that allows for an vertex to have a relationship with itself is called a **pseudograph**. Suppose our graph contains servers that could be infected with malware. If we are modeling the actions of the malware, then an action that affected the server directly would be a relationship with itself.

5.2.4 BIPARTITE GRAPH

A **bipartite** graph is a graph where the vertices can be divided into two disjoint sets such that all edges connect a vertex in one set to a vertex in another set. There are no edges between vertices in the disjoint sets. To illustrate, consider A records and PTR records in DNS. A domain has an A record for an IP address, and an IP address has a PTR record which points back to a domain. A domain does not have an A record for another domain and an IP does not have a PTR record for another IP. So the graph created by combining the A records and the PTR records for domains and IP addresses is a bipartite graph where one set is the domains and the other set is the IP addresses.

A special case of the bipartite graph is the **complete bipartite** graph. In this graph, every vertex of one set is connected to every vertex of another set. We represent a complete bipartite graph by $K_{m,n}$ where m is the size of the first set and n is the size of the second set. So a $K_{3,2}$ complete bipartite graph has three vertices in the first set, two vertices in the second, and every vertex in the second set is connected to the vertices in the first set. We can count the number of edges in this graph, which is 6.

5.2.5 SUBGRAPH

A graph that is contained entirely within another graph is referred to as a **subgraph**. This means that vertices in the subgraph are adjacent if they are adjacent within the larger graph and that they are not adjacent if they are not adjacent in the larger graph. Looking at a subgraph is a way of focusing the analysis on a smaller portion of the graph and sometimes making the problem more tractable.

An **induced** subgraph is created when we take a set of vertices in the graph and create a subgraph of the original graph that contains those vertices and only those vertices. For example, suppose a graph is modeling a BGP peering relationships on the Internet. If we restrict our Autonomous Systems to a particular region, then the induced subgraph is a view of the peering relationships within that region and only that region. We would see how the Autonomous Systems in that region are related but we would see nothing about how that region is connected to the rest of the Internet.

5.2.6 GRAPH COMPLEMENT

The **complement** of a graph is also a graph, albeit a graph created from an existing graph. In this case, we start with a graph G and create a new graph from it. The new graph has the same vertices as the original graph and two vertices in the new graph are connected if they are not connected in the original. In other words, the new graph's relationship is 'not a relationship in the original graph'. We represent the new graph by G^c.

For example, if our graph is modeling connections in a network, the complement of that graph is the entities in the network that do not have connections.

Example 5.2.1. Suppose a party is thrown with six attendees. It can be proven that there are either three mutual acquaintances or three mutual non-acquaintances. In other words, three people at the party are friends or three people at the party are not friends.

This example can be illustrated with a graph to model friendships, where if two people are friends, they are connected by an edge. Proving the statements amounts to showing that either the graph that models friendships or the complement of that graph contains a triangle. This theorem was first proven by a mathematician named Ramsey.

5.3 PROPERTIES OF GRAPHS

We have already learned some graph terminology like vertices, edge, and connections that define the basic parts of graphs. In order to study graphs more completely, we need to expand the vocabulary to include properties of graphs. These properties are found within the graph after it is creation.

5.3.1 GRAPH SIZES

Size is an important property of a graph. Since there are two sets in a graph, the vertices and edges, we have two properties that are related to size. The number of vertices is called the **order** of the graph. In most graphs, the number of edges is much larger than the number of vertices. We call the number of edges the **size** of the graph.

In Section 5.2.4, we defined a complete bipartite graph. For $K_{m,n}$, we know that the order of the graph is $m + n$ and the size of the graph is mn.

5.3.2 VERTICES AND THEIR EDGES

A vertex is called **incident** to an edge if it is part of that edge. Remember that we write the edges as (v_1, v_2), so v_1 is incident to (v_1, v_2) and so is v_2. Two vertices are called **adjacent** if there is an edge between them. In other words, v_1 and v_2 are adjacent. They are also known as **neighbors**.

A vertex with no neighbors is often referred to as **isolated**. This means that the entity represented by the vertex is completely unrelated to the rest of the graph. So if we create a graph with many isolated vertices, we might want to consider the relationships we are modeling with the graph. Clearly some entities in our original set do not possess the relationship we are examining. We can use this to tell us something about our original data set. For example, if we are modeling malicious network connections on a set of IP addresses with a graph, then the isolated IP addresses are those that had no malicious connections.

5.3.3 DEGREE

For any vertex, the number of vertices that are adjacent to it is called the **degree** of the vertex. It is equivalent to the number of neighbors of the vertex. For a vertex v, we let $\deg(v)$ represent the degree.

Example 5.3.1. Studying just the degree can impart interesting information about the data that the graph models. Suppose we have a set of domains and IP addresses with the associated undirected graph in Fig. 5.1. The vertex with the largest degree in the graph is 172.24.125.16 and all of the vertices that are labeled with domains point

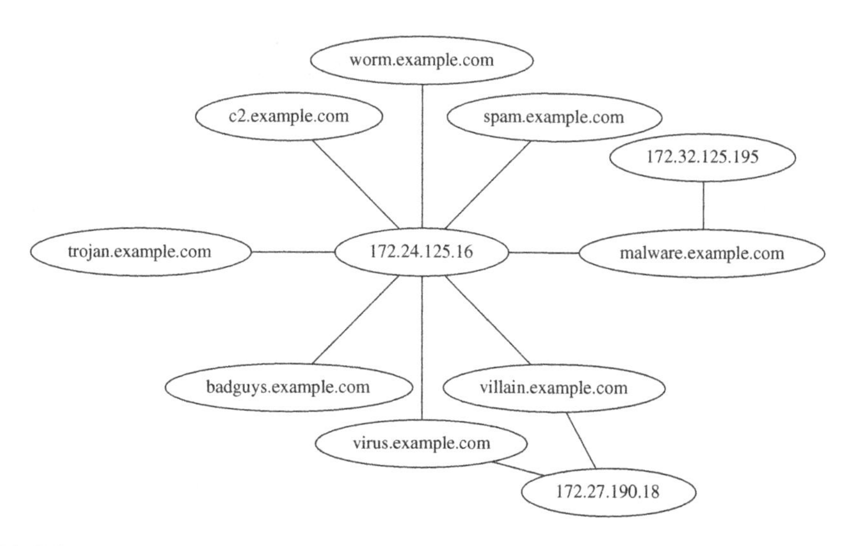

FIG. 5.1

A graph modeling DNS.

to it. If we know that the domains are all related to malware, then that IP address is suspect and should be either blacklisted or monitored.

The degree also has an interesting property. If we sum all of the degrees in the graph, the sum is exactly two times the number of edges. This result is considered one of the first theorems in graph theory, and was proven by the mathematician Leonhard Euler in 1735. It can also be used to show that the number of vertices with an odd degree is even.

5.3.4 DIRECTED GRAPHS AND DEGREES

A vertex in a directed graph can be the origin of an edge to another vertex or it can be the destination. This implies that there are two different measurements of degree. The **indegree** of a vertex is the number of vertices that have relationships with the vertex and pointing to the vertex and the **outdegree** of a vertex is the number of vertices that the vertex has relationships with. The sum of the indegree and the outdegree is the total degree of the vertex.

The indegree is written as $\deg^-(x)$ and the outdegree is written as $\deg^+(x)$. If $\deg^-(x) = 0$, then that vertex has relationships with other vertices but no vertices have relationships with it. We call that a **source**. On the other hand, if $\deg^+(x) = 0$, the vertex has no relationships with any other vertices in the graph and all of its neighbors have relationships with it. This is called a **sink**.

Practically speaking, a source is the origin of relationships within the directed graph and a sink is a destination of relationships within the graph. Finding sinks and sources in directed graphs is very useful for analyzing them. For instance, if we have created a graph using network flow data, a sink is an IP address that does not create any network flow, but where all the network flow goes to it. On the other hand, a source is an IP address that is the origin of the network flow, but never the destination.

5.3.5 SCALE FREE GRAPHS

The distribution of the degrees in a graph, often called the degree sequence, can be informative. If the distribution follows a Zipf distribution, the graph is called a **scale free** graph. The scale free network has several properties, one of which is that vertices with a degree that are much smaller than the average degree of the graph are relatively common. Practically speaking, it means that there are a few vertices with very large degrees and many vertices with rather small degrees.

This does tell us something about the nature of the graph, beyond the fact that there is a power law distribution of degrees. Suppose we are modeling a large network, either using BGP peering information or physical information. If the graph is scale free and failures occur randomly on the vertices of the graph, then there is a smaller chance of the failure affecting a large vertex. This assumes that all failures are random and do not depend on the size of the degree of the vertex.

5.4 PATHS, CYCLES AND TREES

Pick a vertex in a graph at random. Now suppose that that vertex is incident to another vertex in the graph, and repeat that process. Practically speaking, if we traverse each link as we find new vertices, we are taking a walk through the graph. This section will mathematically quantify this process and discuss applications to modeling using it.

5.4.1 PATHS AND CYCLES

A **path** in a graph is essentially a walk through the graph. Mathematically speaking, we write a path as an ordered list of vertices $\{v_1, v_2, \ldots, v_n\}$ where v_i and v_{i+1} are adjacent and each v_i is unique in the path. The length of the path is the number of edges in the path. Think of it as the number of links traversed to get from one vertex to another.

A path can be considered a series of derived relationships. For example, I derive a relationship from my paternal grandparents through my father. I am not directly related to them, but my paternal grandparents are directly related to my father who are directly related to me. So there is a path of length two between my paternal grandparents and me. In a similar fashion, consider peering relationships in the Internet. Suppose ASN65512 is peered with ASN65530, and that ASN65530 is peered with ASN65520. Then ASN65512 has an indirect relationship with ASN65520, in that some of its traffic will come from that ASN65520 or some of the traffic from ASN65512 will go through ASN65520.

A **cycle** is a path that begins and ends at the same vertex. Analyzing cycles is an important part of graph analysis, so we will spend some time covering them in Section 5.4.5.

An interesting path in a graph is called a **Hamilton path**. This is a path that visits every vertex in a graph exactly once. If the path is a cycle, it is called a **Hamilton cycle**. If a graph has a Hamilton path or cycle, we know that every vertex in the graph has a derived relationship with every other vertex in the graph. We also know that if a graph has a Hamilton cycle then it has a Hamilton path, because we can just remove one of the edges from the cycle and create the path.

Example 5.4.1. For a complex example, recall that we said we could model the Autonomous Systems used in BGP as an undirected graph. Routing tables are available at both http://routeviews.org and http://ripe.net and we can use this data to create a graph. A routing table has entries that look like:

$$\text{CIDR ASN}_1, \text{ASN}_2, \ldots, \text{ASN}_n$$

The CIDR block is the network announced by ASN_n and the route is then passed to ASN_{n-1} which then passes it along the chain where the final stop is ASN_1. This means that one path to access the IP addresses in the CIDR block is to start at ASN_1, go to ASN_2 and so on until you reach the home of the CIDR block, which is ASN_n.

The vertices of the graph are then the Autonomous Systems and the edges are derived from the paths in the data set. In other words, ASN_i and ASN_j are connected in the graph if and only if there is a path in the data set that contains ASN_iASN_j or ASN_jASN_i.

From the order of the graph, we know how many Autonomous Systems are active in the Routing Table at that time. The size of the graph tell us how many relationships, or peerings, are active. Each path through the graph to an Autonomous System is a potential path for network traffic to travel through.

Suppose we know that the Autonomous System ASN_q is announcing the network that contains the IP address of our destination. If there is only one path from our starting Autonomous System to ASN_q, then if we lose any peerings on that path we lose our ability to connect to the IP address. On the other hand, if there is more than one path, we have redundancy in accessing the IP address. In routing, it is important to know that whatever path we choose to access the IP address is not necessarily the path the owner of the IP address chooses to return to us. Also, even though the path we find through the graph may not exist in the original data set, we can infer that it would exist due to the graph. The relationships found in the paths exist whether or not we use that exact path.

5.4.2 SHORTEST PATHS

If we have two vertices in a graph and multiple paths between them, then there is a shortest path in that collection. That path is not necessarily unique. For example, consider the graph in Fig. 5.1. There are two paths between 172.24.125.16 and 172.27.190.18 of length 2, each of them the shortest path in this situation. The shortest path problem is the process of finding the shortest path between two vertices on a graph. We can consider it the most efficient route through the graph.

Another way of considering the shortest path problem is to remember that a path is a series of derived relationships. The shortest path is the series with the shortest derivation, or the closest relationship. Since a graph is modeling relationships, we are often interested in the closest relationships.

There are many algorithms that are used to find the shortest path in a graph, for example, Dijkstra's algorithm is a popular algorithm. The implementation of this algorithm is not of interest to us, rather, we are more interested in what the shortest path in a graph means.

5.4.3 CONNECTED AND DISCONNECTED GRAPHS

An undirected graph where every vertex is connected to every other vertex by a path is called a **connected graph**. A disconnected subgraph is a connected subgraph of the original graph that is not connected to the original graph at all. A **component** of a graph is a disconnected subgraph. For example, if we have a social network with three components, then we have three groups of friends who have no common friends. Similarly, if we are examining malicious domains and IP addresses, a component is a set of relationships between the domains and IP addresses that is not related to any other malicious domain in the original set.

A good first step in graph analysis is to find the components in the graph. This is to ensure that we are looking at interrelations, not impossibilities. For example, if we are looking at a social network, and Alice and Bob are not connected by a path in the graph, then trying to analyze the nonexistent relationship between Alice and Bob is a waste of time.

5.4.4 TREES

A graph that has no cycles is called a **tree**. An important thing we know about trees is that the number of edges minus the number of vertices plus the number of components is always equal to zero. If there is just one component, then the number of edges equals the number of vertices plus one.

It is also true that every two vertices in the tree are connected by a unique path. This is easy to see. Suppose there were two different paths that connect two different vertices in the tree. Then a cycle can be constructed in the graph with the two paths, and it is not a tree any more.

The fact that a graph is a tree can tell us interesting things about the data we are modeling. One thing is that there is no redundancy in the relationships between data in the vertices. For example, if we are looking at routing data, it means that there is only one path between any two Autonomous Systems. It also means that if an Autonomous System is removed from the graph, then the graph is going to be split into two components.

5.4.5 CYCLES AND THEIR PROPERTIES

The presence of a cycle within a graph implies redundancy within the relationships. If a physical network is modeled with a graph, then having a cycle within it means that if we lose a physical link within the network, we have not lost the entire network. Similarly, if we are modeling a routing graph and that has a cycle, if we lose a peering relationship, then we have not lost our entire BGP peering network.

The shortest cycle possible in a graph is a cycle of length three, that is, a triangle. This is an interesting study of relationships because it says that two entities who have a relationship have a direct relationship with each other as well. If we are considering social networks, it means that two of my friends are also friends, so it is a tight cluster. We examine triangles further in Section 5.7.

What is the absolute minimum number of cycles available in the graph? This value goes by many names. It is called the **fundamental number of cycles**. the **Betti number** of the graph, the **Cyclomatic number** of the graph, or the **Cycle Rank** of the graph. It can be represented by multiple symbols. Two of the more common are $\beta(G)$ or $m(G)$. We will use $\beta(G)$ and call it the fundamental number of cycles.

The fundamental number of cycles is actually easy to compute. Let $|E|$ be the number of edges in the graph, $|V|$ the number of vertices and $|C|$ be the number of components. Then it is given by Eq. (5.1):

$$\beta(G) = |E| - |V| + |C| \tag{5.1}$$

As we stated previously, the fundamental number of cycles in a tree is 0.

5.4.6 SPANNING TREES

Every connected graph contains a subgraph that is a tree. Since a single edge is effectively a tree, then this can be considered a somewhat simple statement. A more complex tree is called a **spanning tree**. This is a subgraph of a graph that touches every vertex and is a tree. There is generally more than one spanning tree in a graph, except in the case where the original graph is a tree.

Another property of the spanning tree is that if you add one more edge to it, you no longer have a tree. You have found a cycle. There is actually a relationship between the fundamental number of cycles and the maximal spanning tree. When a spanning tree of a graph is created, there are edges of the original graph that are not included. The exact number of edges left over is the fundamental number of cycles.

5.5 VARIETIES OF GRAPHS REVISITED

Using the properties defined in previous sections we can now define more kinds of graphs. In these cases, the kind of graph is not determined until after the modeling is complete. In other words, we take our data, create the graph, and then examine the graph afterwards using the techniques found in this section. This will yield more information about the structure of our graph.

5.5.1 GRAPH DENSITY, SPARSE AND DENSE GRAPHS

The maximum number of edges that a graph with n vertices can have is $\frac{1}{2}n(n-1)$. A graph that has close to this number of edges is called a **dense** graph. At the other extreme is a graph with only a few edges, called a **sparse** graph. The measure that quantifies this is called the **graph density**, given by Eq. (5.2), where $|E|$ is the number of edges and $|V|$ is the number of vertices:

$$D(G) = \frac{2|E|}{|V|(|V|-1)} \tag{5.2}$$

Graph density really is defined as the total number of edges in the graph divided by the total number of edges that could be in the graph. If the density is 0.5, then it has half of the maximum number of edges. If the graph has no edges, meaning it is just a series of vertices with no relationships, then the density is 0. Returning to the complete bipartite graph $K_{m,n}$, the graph density of it is given by $\frac{2mn}{(m+n)(m+n-1)}$.

Let us look at the complement of a graph again. The edges of the complement are the edges that are not in the original graph, so the number of edges in the complement is the total number of edges there could be minus the number of edges in the original graph. After a bit of algebra, we have Eq. (5.3):

$$D(G^C) = 1 - D(G) \tag{5.3}$$

This implies that if G is a dense graph, then G^c is a sparse graph and vice versa.

5.5.2 COMPLETE AND REGULAR GRAPHS

If every vertex on the graph is connected to every other vertex on the graph, the graph is called a **complete** graph. This means that every entity on the graph has a relationship to every other entity on the graph. A triangle between three vertices is the simplest example of a complete graph. Another way to think of a complete graph is that the degree of each vertex is one less than the number of vertices. We represent complete graphs on n vertices as K_n.

We know that if there are n vertices in a complete graph, then there are $\frac{1}{2}n(n-1)$ edges in the graph. In other words, a complete graph is dense.

A **k-regular graph** is a graph where the degree at each vertex is equal to k, also known as a **regular graph of degree** k. So a complete graph K_n is a regular graph of degree $n-1$ or an $n-1$ regular graph. A 1-regular graph is just a series of two vertices connected by one edge, and those vertices are not connected to any other edges in the graph. It has been proven that for a set of n vertices, a k-regular graph can only exist if $n \geq k+1$ and nk is an even number. So if we have nine vertices, then it is impossible to draw a 3-regular graph on it.

The number of edges in a k-regular graph with n vertices is $\frac{1}{2}nk$. So we know that a 3-regular graph on six vertices has 9 edges. We could also draw that out and count the edges, but luckily for us we do not have to.

5.5.3 WEIGHTED GRAPH

A **weighted** graph is a graph whose edges are assigned values, known as weights. For example, if we are modeling network flow, then the vertices could represent source and destination IP addresses, while the weight could be the length of time that a flow lasted, or the number of bytes that was transferred during the flow.

The total weight of a graph or a subgraph is the sum of the weights on the entire graph or subgraph. We can consider every graph a weighted graph if we set the weights to be 1 on every edge. Then the weight of the graph is the number of edges.

The weighted graph has implications for the shortest path problem discussed in Section 5.4.1. The shortest path problem now involves finding the path with the combination of fewest links and least total weight. This means that the shortest path in a weighted graph could actually have more links in it than the same graph without the weights.

If we consider BGP routing again, suppose we want to ensure that our out bound traffic does not cross a certain provider's network. To ensure this, we would assign a very high value to each edge that has an incidence with that provider's ASNs. Then using the shortest path problem, we can ensure that we choose other paths.

5.5.4 AND YET MORE GRAPHS!

There are many other kinds of graphs, that we will touch on briefly. A **cycle graph** is a graph where the entire graph consists of a single cycle, that is, the fundamental number of cycles is 1 and every vertex on the graph has degree two. It is also a simple example of a Hamilton graph.

There is also the **trivial** graph, which is a graph with all vertices and no edges. In each case, at first glance it seems we are not necessarily modeling anything interesting or useful, but even the trivial graph could indicate a significant result. When modeling a virus infecting a network, the trivial subgraph of that infection would be the systems that are not affected. On the other hand, if our calculations yield the trivial graph, we may have some assumptions that need to be readdressed.

5.6 REPRESENTING GRAPHS

Visualizing a graph is a useful but time consuming process that should not be done by hand. We want to automate this as much as possible. To do this, we have to take our graph and turn it into something the computer will understand so the computer can draw our graph for us. Computers want numbers and they do understand matrices. That combination will yield a nice representation that can be used by a computer to draw the graph as well as analyze it.

5.6.1 ADJACENCY MATRIX

One of the most common representations of a graph for computer processing is called an **adjacency matrix**. In the adjacency matrix, the rows and columns are each defined to be the vertices in the graph. Let us call them $\{v_1, \ldots, v_n\}$. In our matrix A, we will put a one at the ith row and jth column if there is an edge between v_i and v_j. Otherwise, we put a zero.

The matrix is an adjacency matrix because we used adjacencies in our graph to create it.

Example 5.6.1. We are going to look at BGP peering, as illustrated in Fig. 5.2.
To create the adjacency matrix, we begin by ordering the vertices.

1. ASN65512
2. ASN65517
3. ASN65520
4. ASN65524
5. ASN65527

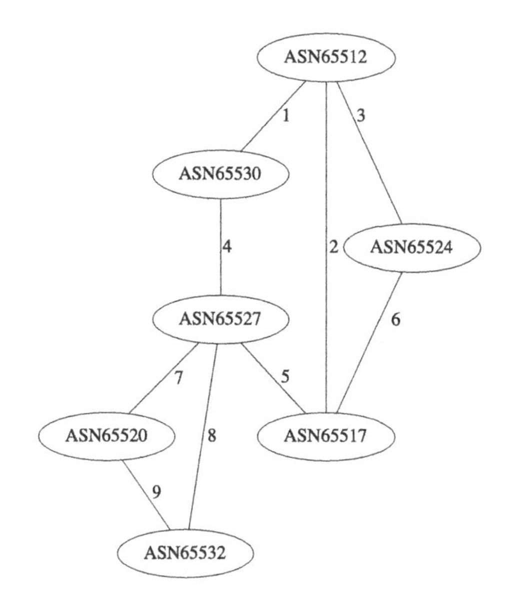

FIG. 5.2

A graph modeling BGP peering relationships.

6. ASN65530
7. ASN65532

Using this ordering and the graph in Fig. 5.2, we can create the matrix in Eq. (5.4). The element in the first row and second column is 1 because there is an edge between ASN65512 and ASN65517:

	ASN65512	ASN65517	ASN65520	ASN65524	ASN65527	ASN65530	ASN65532
ASN65512	0	1	0	1	0	1	0
ASN65517	1	0	0	1	1	0	0
ASN65520	0	0	0	0	1	0	1
ASN65524	1	1	0	0	0	0	0
ASN65527	0	1	1	0	0	1	1
ASN65530	1	0	0	0	1	0	0
ASN65532	0	0	1	0	1	0	0

(5.4)

The adjacency matrix for an undirected graph has a nice property, namely, it is symmetric. This means the value in the *i*th row and *j*th column is the same as the value in the *j*th row and *i*th column. Another interesting fact involves matrix multiplication. If we multiply the adjacency matrix with itself, then if there is a nonzero value in the *i*th row and *j*th column, there is a path from v_i to v_j of length two. It does not tell us what the path is, just that there is a path. The nonzero value also tells us how many distinct paths there are.

If we multiply it by itself again, for a total of three times, we see if there are any paths of length 3. In this case, if there is a nonzero element in the diagonal of the graph, that is, the ith row and column, then there exists a cycle. You can multiply the graph by itself as many times as you want to find paths of that length. Remember though, this does not tell you what the paths are, just that they exist and how many unique ones there are. Finding them is another problem.

5.6.2 INCIDENCE MATRIX

Another matrix used to describe a graph is the **incidence matrix**. This uses the incidences between edges and vertices to create the matrix.

We begin by ordering the set of edges $\{e_1, e_2, \ldots, e_n\}$ and then ordering the set of vertices $\{v_1, v_2, \ldots, v_m\}$. The matrix M in its ith row and jth column has the number of times the jth edge is incident with the ith vertex.

Example 5.6.2. We return to Fig. 5.2 used in Example 5.4. The edges are labeled and ordered and we can use the ordering of the vertices used in that example.

Now we can create the incidence matrix. The columns are the edges and the rows are the vertices:

	1	2	3	4	5	6	7	8	9
ASN65512	1	1	1	0	0	0	0	0	0
ASN65517	0	0	0	0	1	1	0	0	0
ASN65520	0	0	0	0	0	0	1	1	1
ASN65524	0	0	1	0	0	1	0	0	0
ASN65527	0	0	0	1	1	0	1	1	0
ASN65530	1	0	0	1	0	0	0	0	0
ASN65532	0	0	0	0	0	0	0	1	1

(5.5)

Most graphs have more edges than vertices, so it is more common that an adjacency matrix is used to represent the graph rather than the incidence matrix. On the other hand, if you have the incidence matrix you can actually derive the adjacency matrix with a bit of linear algebra and vice versa, so it does not really matter which form you choose.

5.7 TRIANGLES, THE SMALLEST CYCLE

5.7.1 INTRODUCTION AND COUNTING

As we said in Section 5.4.5, triangles are the smallest cycle. They imply a tightness in the relationship between the three vertices. If the graph is modeling relationships, finding these tight groups is useful.

One thing we know about triangles is that for a graph with n vertices, if there are $\frac{n^2}{4} + 1$ edges, there must be a triangle. In other words, the maximum number of edges in the graph with no triangle at all is $\frac{n^2}{4}$. This is not saying that there cannot be a triangle in a graph with fewer than that number of edges. It is just saying that if there are more than that number of edges, there must be a triangle.

For counting the number of triangles in a graph, recall that in Section 5.6.1, we said that if adjacency matrix is multiplied with itself three times, and the ith row and column is nonempty, then there is a cycle of length three. We also know how many paths there are from the ith vertex to itself. We will represent the element of this matrix in the ith row and column by $v_{i,i}$. So that means, if we consider the fact that there are two ways to go around a triangle, there are exactly $\frac{1}{2}v_{i,i}$ triangles containing that vertex.

We can go one step further and find out how many triangles are in the graph. If there are $\frac{1}{2}v_{i,i}$ triangles for every ith vertex, then we can just sum that up for every vertex. If we do this, then we are counting each triangle three times. Once for the first vertex in it, then another time for the second vertex in it, and finally for the third vertex. So we solve that by multiplying our result by $\frac{1}{3}$. Written mathematically, it looks like Eq. (5.6):

$$\frac{1}{6}\sum_{i=1}^{n} v_{i,i} \tag{5.6}$$

With the graphs illustrated in this chapter, like Fig. 5.2, we can look at it and count the number of triangles. Two are present in that graph. But once the graph gets more complex, using a bit of matrix multiplication and Eq. (5.6) allows us to determine the number of triangles in the graph easily.

5.7.2 TRIANGLE FREE GRAPHS

For the converse of the previous discussion, we turn to the question of triangle free graphs, that is, graphs with no triangles at all. A triangle free graph means that there is no close group of relationships found within the graph, that things are more spread out. A bipartite graph is an obvious example of a triangle free graph.

Remember that a bipartite graph, as defined in Section 5.2.4, is a graph where we can divide the vertices into two sets. Each edge in the bipartite graph has a vertex from one of the two sets, but the edge is not incident to vertices in the same set. So there are no triangles in a bipartite graph.

In Section 5.7.1, we said that for a graph with n vertices, if there are more than $\frac{n^2}{4}$ edges in it, it has a triangle. This actually gives us a maximum for the number of edges in a bipartite graph. In fact, it has been proven that the biggest triangle-free graph with n vertices is a complete bipartite graph.

5.7.3 THE LOCAL CLUSTERING COEFFICIENT

We begin by defining a **neighborhood** of a vertex. This is where we take the vertices that are neighbors of a given vertex and create the induced subgraph on

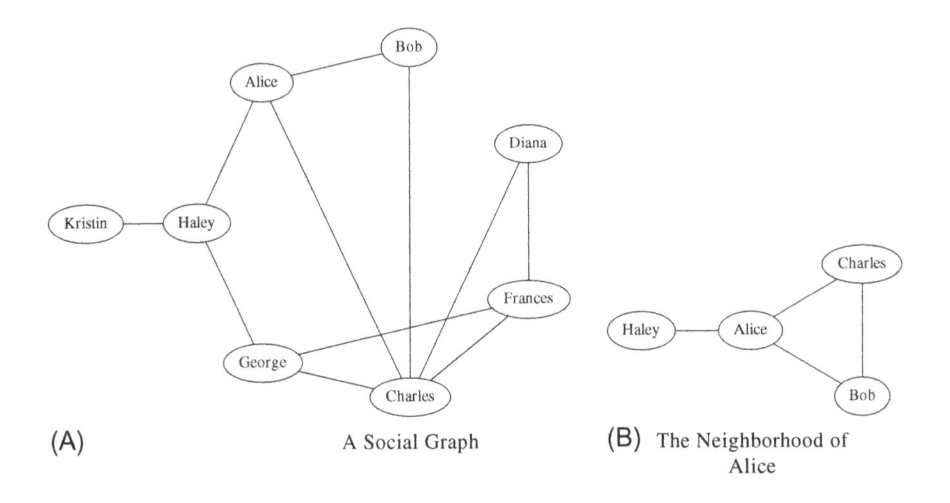

FIG. 5.3

A graph and a neighborhood. (A) A social graph. (B) The neighborhood of Alice.

this set, including the original vertex. If v is the vertex of the graph, we represent the neighborhood as N_v.

In Fig. 5.3A, we have modeled a social network. In Fig. 5.3B, we have taken Alice and displayed her neighborhood. In essence, the neighborhood is the vertices adjacent to Alice, which are Haley, Charles and Bob. Then we also add the edge between Charles and Bob. We do not add the edge between Haley and Kristin, as Kristin is not adjacent to Alice. Another way of describing this is by saying that Alice is friends with Haley, Charles and Bob, and Bob and Charles are friends. However, Haley has no relationship with Charles or Bob.

The neighborhood has one triangle in it. On the other hand, if Charles and Haley were friends, it would have two triangles in it. The **Local Clustering Coefficient of a vertex** is the ratio of the number of triangles in the neighborhood of the vertex divided by the maximum number of triangles that could be in the neighborhood. Another way of describing it is to say it is the number of my friends that are also friends divided by the total number of my friends that could be friends.

Mathematically speaking, suppose our vertex has degree d. If T is the number of triangles in the neighborhood, then the Local Clustering Coefficient, represented by C_v, is given by Eq. (5.7):

$$C_v = \frac{2T}{d(d-1)} \tag{5.7}$$

Another way to look at it uses the fundamental number of cycles of the neighborhood. Then the Local Clustering Coefficient of a vertex is given by Eq. (5.8):

$$C_v = \frac{\beta(N_v)}{\beta(K_{d+1})} \tag{5.8}$$

The Local Clustering Coefficient of a vertex can be considered a measure of how close the neighborhood of that vertex is to being a complete graph. If it is 0, then the neighborhood of a vertex is a tree. If it is 1, then the neighborhood is a complete graph. Returning to Fig. 5.3A, then $C_{\text{Bob}} = 1$ and $C_{\text{Haley}} = 0$. On the other hand, $C_{\text{Charles}} = \frac{1}{5}$.

The **Local Clustering Coefficient of a graph** is the average of the Local Clustering Coefficients of all of the vertices in the graph. The Local Clustering Coefficient of the graph in Fig. 5.3A is 0.454.

5.8 DISTANCES ON GRAPHS

The distance between two vertices in a connected graph is related to the paths between the vertices. We know that in a connected graph, every two vertices have at least one path between them. From Section 5.4.1 we know that the length of a path is the number of edges in the path. So we could say that the distance between two vertices is the length of a path between them. The question becomes which path should we choose. For consistency, we will use the length of the shortest path. Of course, there can be more than one path that has the shortest length, but that will not affect the distance.

The distance between two vertices using the shortest path distance will be written $d(u, v)$. We can use this value to find additional attributes of the graph that are related to the structure.

5.8.1 ECCENTRICITY

The **eccentricity** of a vertex is the maximum of all the distances between it and every other vertex in the graph. That is, the longest shortest path between a vertex and every other vertex in the graph. If we think of the underlying data, it is the furthest distance between two data points being modeled by relationships.

The **radius** of a graph, written $r(G)$, is the minimum of all of the eccentricities in the graph. In other words, it is the closest furthest distance between two vertices. The **diameter**, written $d(G)$, is the maximum of the eccentricities. A companion to the radius, it is the furthest distance between two vertices. In a circle, we know that a radius is half of a diameter. The same is not exactly true in this case. In this case, we can show that the inequalities in Eq. (5.9) are true:

$$r(G) \leq d(G) \leq 2r(G) \tag{5.9}$$

If the eccentricity of a point is equal to the radius of the graph, we call that a **central point**. The **center** of a graph is the set of all central points. A **peripheral** vertex is a vertex whose eccentricity is equal to the diameter of a graph. It is possible for all the vertices in a graph to be in the center. Suppose our graph is the regular graph on three vertices, or K_3. Then eccentricity of each vertex is 2, so that $r(K_3) = d(K_3)$. This implies that every vertex in K_3 is both a central point and a peripheral vertex.

5.8.2 CYCLE LENGTH PROPERTIES

The length of the shortest cycle in the graph is called the **girth**. If a graph contains a triangle, then the girth of the graph is three. On the other hand, if the graph is a tree, the girth is defined to be infinity. So the girth is an integer that can be anywhere between three and infinity. If the girth of a graph is four, then the graph is a triangle free network. You cannot argue it the other way, though. If the graph is a triangle free network, then you can say the girth is greater than three. You cannot say that it is four without further examination of the graph.

Girth is important in determining an optimum number of sensors to be placed in a network so that the sensors are not detected. The larger the girth, the easier it is to determine this number. A graph with a large girth has also been studied in the spread of viruses. If the smallest cycle in the graph is large, then it is easier to isolate infected systems before they are infected. A girth of three in this case means that if two vertices in the cycle are represent infected systems, a system represented by the third vertex in the cycle is most likely to be infected before it can be isolated, as it can be infected by either system.

The length of the longest cycle is the **circumference** of the graph. If the graph is a tree, we define it to be infinity, as in the case of the girth. It is the longest walk you can take on a graph and end up back where you started from without crossing your path on your walk. It can be equal to the girth of the graph, which tells you something about the graph itself. It tells you that every cycle in the graph is no shorter than the girth and no larger than the girth. In other words, every cycle in the graph is exactly the length of the girth.

5.9 MORE PROPERTIES OF GRAPHS

As in Section 5.3, this section will cover additional properties found in graphs. These properties yield information about the graph itself and in general will cover ways to understand the graph in meaningful ways. If we are modeling a computer network, we want to know what edges would cause our network to lose connectivity. If we are modeling a social network, or a threat actor network, we want to know what people are crucial to the network and would cause it to fall apart otherwise.

5.9.1 CUT

A **cut** is a vertex in a graph that, when removed, separates the graph into two nonconnected subgraphs. It is possible that if we remove the vertex, we are left with one subgraph consisting of a single vertex and a large graph, in which case we call the cut point **trivial**. If our graph is a tree, we know that every vertex in the graph is a cut point.

If v is a cut of a graph G, then we know we can find two more vertices w and x of G where v is on every path between w and v. We know this because a graph is disconnected if there are two vertices in the graph that do not have paths between them. The cut is then the vertex that allowed the path between the two vertices and by removing it, we have removed the paths. A cut is also a way of finding a vulnerability in the graph. Suppose we are modeling the topology of a company's network using a graph. If we find a cut in the graph, then we have found a node in the infrastructure that could cause a disruption to traffic if it goes down. A stable and redundant network topology does not have any cuts that could disrupt the entire network.

If there are no cut points on a graph, it is called **nonseparable**. The largest nonseparable subgraph of a graph is called a **block**. Blocks are interesting because even if you remove a vertex from the graph, there is still a path between any two vertices. In other words, there is at least two distinct paths between any two vertices, which means that any two vertices in the graph lie on a cycle. Also, if we consider three vertices in the graph, for any two of them there is a path that does not contain the third. If this was not true, then we would have for any two vertices all paths that go between them contain the third vertex, and this would make the third vertex a cut.

In other words, the presence of a cut and the lack of presence of a cut tell us quite a bit about the structure of the graph in terms of redundancy and stability.

5.9.2 BRIDGE

The cut referred to a vertex that separated a graph, a **bridge** refers to the edge whose removal will separate the graph into two disconnected subgraphs. We know that if e is a bridge in G, then e cannot be on any cycle of G. Another thing we know is that the bridge breaks up the connectivity between at least two vertices. This means that there are two vertices in the graph where the edge e is on every path between the two vertices.

In other words, this relationship is part of a derived relationship between two entities represented in the graph. In social networking terms, a bridge is a relationship between two separate groups of people. For example, if we have a social network of people who write malware and another social network of people who buy malware to use it, then a bridge between the two groups would be a single relationship between a maker of the malware and a buyer of the malware. Finding this relationship and breaking it would mean that those who make malware would not have an outlet to sell malware until another relationship was formed.

5.9.3 PARTITIONS

In a bridge, the edge that is removed leaves behind the two disconnected subgraphs. The subgraphs are of arbitrary size. It is possible that removing the edge leaves behind a single edge and a large graph, or in other words, two subgraphs of very different sizes. A **partition**, also know as the **edge separators**, of a graph is when one or more edges are removed to create two separate and equal sized connected subgraphs. This is a difficult problem, so algorithms exist to find approximations. We find two separate but close to equal size connected subgraphs in a partition.

In the case of a weighted graph, a partition is the creation of two subgraphs of approximate equal weight. So if we have a graph that is weighted heavily on just a few edges but lightly on the rest, then the sizes of the subgraphs resulting from a partition might be very different.

For a simple example, suppose our graph models malware authors and malware vendors. We weight the malware author very high and the vendors relatively low. Finding the edges in the partition of the graph modeling the relationships between the two groups would tell us what relationships we need to sever.

A ***k*-partition** is when the graph is partitioned into k similar sized pieces. A network defender into how many pieces to partition their network in order to yield similar size connected networks that can then examined for evidence of malicious network activity, rather than trying to attack the entire problem at once. The size of k is determined partly by the network defender and partly by the size of the graph. You want to reduce your workload, so you choose k so that it does so in a meaningful way.

5.9.4 VERTEX SEPARATORS

As a cut is the vertex version of a bridge, **vertex separators** are the vertex version of the partition. For vertex separators, we look for the vertices in a graph that, when removed, create two disconnected and equal sized subgraphs. As in the partition case, it is difficult for an algorithm to find exactly equal subgraphs and we look for approximate solutions.

A difference between the partitions and vertex separators is that when you remove an edge, you only break one connection. Removing a vertex breaks every connection of which that vertex is a part.

As in the partition case, if the graph is weighted then we are interested in similar total weights for the two subgraphs, not necessarily two subgraph of the same size. Also, you can find the ***k*-vertex separators** to separate the graph into k pieces each of either the same size, or in the case of a weighted graph, the same total weight.

The vertex separators in a routing graph are those Autonomous Systems whose removal would break the network into two distinct pieces. In other words, if you wanted to break the Internet into two equal halves, these are the Autonomous Systems to remove.

Suppose a cellular network is infected with a worm. Finding the vertex separators of the network allows us to break the network into k pieces and clean each individual piece. This approach allows communication within the partition, but prevents outside communication; specifically, worm propagation.

5.9.5 CLIQUES

In a complete graph, every vertex in the graph has a relationship with every other vertex in the graph. For example, in a complete social network, every person in the network would have a relationship with every other person in the network. Not every graph is a complete graph, but many contain one as a subgraph. We call that subgraph a **clique**. The size of largest clique within a graph is called the **clique number** of the graph and is generally denoted $w(g)$. A famous application of cliques is determining the capacity of the trunk of a car. Cliques can also be used in wireless security and to analyze communication networks in general.

If a graph is a bipartite graph, there are no cliques within it, as one set of vertices only has connections to the other set of vertices and there are no connections within the sets of vertices. On the other hand, if the clique number of a graph is three, then the largest complete subgraph in the graph is a triangle. Depending on the size of the graph, the graph could be spread out. The clique number is useful for determining structural information about the graph.

5.10 CENTRALITY

The idea in centrality is to identify important vertices in a graph. These nodes can be key when searching for the origin of a virus infection and also can be useful in studying a social network. Suppose we are using a graph to model the social interactions of malicious software authors. Using one of the methods defined in this section, we find a central vertex. This vertex could be a facilitator for the malware authors or an originator for new software that spreads through the network. Without further study of the central vertex, we do not know. The point, however, in this section, is the ability to find that central vertex.

There is not really a qualitative measurement available for centrality, it is more often a comparison. Also, one centrality measure does not fit all situations. A measure related to cliques would not apply in a graph that does not contain any cliques.

One thing to note, however, we will assume every graph in this section is connected. It is hard to consider a vertex important when there are vertices it has no relationship to, so for this section, all graphs are connected.

5.10.1 BETWEENNESS

There is an ancient proverb "All roads lead to Rome." If we assume this is true in the simplest reading of the proverb, then to travel from Paris to Athens we must go through Rome. In other words, Rome is between every other two cities. For a graph, we want to find the vertex equivalence of Rome. That is, a vertex that is between all other vertices when we consider the shortest path between vertices, and if not, how close can we get. We begin by defining a **betweenness measure**.

Let $\sigma_{(u,v)}$ be the number of shortest paths between the vertices u and v. Now let $\sigma_{(u,v)}(w)$ be the number of shortest paths between u and v that pass through the vertex w. Then the betweenness of a vertex is given in Eq. (5.10):

$$g(w) = \sum_{u,v \in G} \frac{\sigma_{(u,v)}(w)}{\sigma_{(u,v)}} \tag{5.10}$$

So it is the sum of the ratio of the number paths through a given vertex divided by the number of paths that exist. If the ratio $\frac{\sigma_{(u,v)}(w)}{\sigma_{(u,v)}}$ is 1, then all of the shortest paths between u and v contain w. If it is 0, then none of them contain w. If the sum of all of the ratios is $|V| - 1$ where $|V|$ is the number of vertices, then every shortest path between two vertices must pass through w. In other words, we have found Rome.

5.10.2 DEGREE CENTRALITY

Another centrality measure, called the **degree centrality**, is based on the degrees in the graph. It can be summarized by "He with the most toys, wins." In other words, the number of neighbors a vertex has is important. If the network is spread out, then there should be low centralization. If the centralization is high, then vertices with large degrees should dominate the graph.

We are going to use Freeman's general formula to compute degree centralization. First, find the vertex with the highest degree. We will call it v_*. Then we will define $H = (|V| - 1)(|V| - 2)$. The degree centrality is defined in Eq. (5.11):

$$C_D(G) = \sum_{v \in G} \frac{|\deg(v_*) - \deg(v)|}{|H|} \tag{5.11}$$

In the numerator of the sum, we are considering how different the degrees are between the largest degree and the given vertex. In the denominator, we are normalizing the results. If the degree of a vertex is the same as the highest degree, then we contribute nothing to the sum. The closer to the highest degree, the less we add to the total sum.

5.10.3 CLOSENESS AND FARNESS

The concepts of **closeness** centrality and **farness** centrality are closely related. The closeness is defined so that if a vertex is close to every other vertex, then the value is larger than if the vertex is not close to everything else. Closeness of a vertex is defined in Eq. (5.12):

$$C(v) = \sum_{w \in G} \frac{1}{d(v, w)} \tag{5.12}$$

In other words, if the sum of the distances is large, then the closeness is small and vice versa. A vertex with a high closeness centrality would mean it has close relationships with many vertices.

Farness centrality is the reciprocal of the closeness centrality so that if the closeness is small, then the farness is large and vice versa. In short, it is the sum of all the distances from the vertex v to every other vertex in the graph. If it is large, then the vertex does not have close relationships with many vertices. That vertex is connected to the graph but far from most other vertices in the graph.

5.10.4 CROSS-CLIQUE CENTRALITY

In high school, a student can be a member of the student council, the chess club, the computer club, habitat for humanity, band, and debate club. If a norovirus were found to be infecting the school, then that student is highly likely to spread that virus among many groups. In other words, if a student is a member of multiple cliques in the school, then the student is a prime candidate for centrality, as defined by cliques.

This discussion leads us to the **cross-clique** centrality, which measures how popular a vertex is among cliques found in a graph. Cross-clique centrality is measured by counting the number of cliques the vertex occurs in. The more cliques to which a vertex belongs, then the more central the vertex. This is a useful measure if you are looking at virus or worm propagation.

5.11 COVERING

In mathematics, when we create a **cover** of a mathematical construct, we are creating a simpler version of the construct that maintains some, but not all, information about the construct. In graph theory, a graph cover is a subgraph of the graph that maintains certain information that has been determined, but is simpler to work with than the entire graph. It is often difficult to find the most optimum cover of a graph.

5.11.1 VERTEX COVERING

First we define the **vertex cover**. In this case, we look for vertices that maintain information about the graph, but are not necessarily the entire set. The vertex cover is the set of vertices where every edge of the graph is incident to at least one member of the set. We can take the vertex cover to be the entire set of vertices in the graph, but in general, we do look for a smaller set.

The smallest set of vertices that fits the definition of vertex cover is called the **minimum vertex cover**. The number of vertices in this set is the **vertex covering number**. A vertex cover is a useful set. It can be used to determine sensor for sniffing network traffic, if we want to ensure we are covering every link within the network graph. We place a sensor at each vertex in the vertex cover and we know that every edge within the network is attached to a sensor. For a graph of large size, a vertex cover can be very hard to compute. We generally look for approximations of the set.

A complementary idea to the vertex cover is the **independent set of vertices**. Vertices are in the independent set if they are not adjacent to each other. A single vertex is an independent set, as it is not adjacent to itself, the only other member of the set. The other way to consider the independent set of vertices is that it is the set of vertices that do not have direct relationships with each other. If our graph is a genealogy tree, then the grandparents and the grandchildren are an independent set of vertices. The reason it is a complementary idea of the vertex cover is that the complement of a vertex cover is an independent set of vertices.

5.11.2 EDGE COVER

An idea related to the vertex cover is the **edge cover**. In this case, we are looking for a set of edges to simplify our graph. In other words, a subset of the original relationships that created the graph. To find this, we find a set of edges where each vertex of the graph is incident to at least one edge in the set. The entire set of edges for a graph is an edge cover, but as in the case with the vertex cover, we generally look for a smaller set.

The set of the minimum number of edges that cover a graph is called the **minimum edge covering**. The size of this set is called the **edge covering number**. For example, suppose our graph is regular on three vertices, in other words, a triangle. Then a minimum edge cover of this graph would consist of two out of the three edges in the graph. It does not matter which two edges we pick, they each cover the graph. So if we find a minimal edge cover of a graph, there is no rule that it must be unique.

The edge cover has been shown to be of use in slowing, or even stopping, a virus spread through a network. We begin by modeling the internal network as a graph, finding the minimum edge cover, and disconnecting those edges in the network. Sensor placement has also been modeled as an edge cover problem.

5.12 CREATING NEW GRAPHS FROM OLD

We use graphs to model a set of data that contains relationships. What happens if we have multiple data sets? In set theory, we can intersect them, union them, take the cross product, and more. These ideas can easily be extended to graphs. For example, if we have a graph created from network traffic on Monday and another graph created from network traffic on a Tuesday, a common question is "how do they differ?"

Another question is "how are they similar?" In a manner similar to set theory, we can take our original graphs and create new graphs containing information about both of the graphs.

5.12.1 UNION GRAPHS

We can combine two graphs into one by taking the **union** of the graphs. If $G_1 = (V_1, E_1)$ is one graph and $G_2 = (V_2, E_2)$ is another graph, then we can combine the graphs into one by first taking the union of the vertices, $V_1 \cup V_2$. Then there is an edge in the new graph if it is an edge in the original graph. If the original two graphs are connected, it does not mean the new graph is connected. If they share no vertices in common, then the union will not be connected.

It does not make much sense to create the union of two graphs that model completely different things. If one graph is modeling the Autonomous System graph of the Internet and the other is modeling DNS resolutions, then taking the union of the two graphs will not help in the analysis of the data. However, if we are considering 2 days of network flow data or two graphs modeling DNS resolutions then we can union the two graphs and discover patterns over both data sets.

There is also no rule that says you can only take the union of two graphs. You can also take the union if you have at least two or more graphs. After a week of traffic collection, suppose we have a graph for each day of the week that models the traffic on those days. Now we can create the graph for the entire week by taking the union of all of the graphs.

5.12.2 INTERSECTION GRAPHS

Another method to combine two graphs into one is by taking the **intersection** of the graphs. If we have two graphs, $G_1 = (V_1, E_1)$ and $G_2 = (V_2, E_2)$ then the vertices of the new graph is the set $V_1 \cap V_2$. Then the edges of the new graph are those edges which appear in both graphs. It is entirely possible that there are no vertices that appear in both graphs. In this case, the two graphs share nothing in common. It is also possible that while they share vertices, they do not have edges in common. Both of these cases tell us something about the graphs we started with and their relationships to each other. In the first case, we know they have nothing in common and no relationship. In the second case, we know they were modeled using some of the same entities, but with completely different relationships. For example, if one graph is about the traffic between IP addresses and the other graph is the resolution of those IP addresses to domains, they would share the IP addresses in common but no relationships.

As with unions, there is no rule that says you can only take the intersection of two graphs. You can do any number of graphs, as long as there is at least two. Returning to our example of traffic collection, if we take the network traffic graphs for a week and then find the intersection over all of the graphs, what we find is the IP addresses that talked to each other every day of that week.

5.12.3 UNITING GRAPHS

In the previous two sections, the idea of combining two graphs that model the same concepts has been discussed. Suppose we have two graphs that model somewhat disparate concepts. For example, one graph models the IP resolutions in DNS and one graph models the domain names accessed in a corporation, uniting the two graphs into one can make analysis much easier. Rather than trying to analyze the separate graphs and uniting the results, we unite the graphs into one and analyze that.

The first method we discuss is called the **join**, also known as the **complete join**. We begin with two graphs, $G_1 = (V_1, E_1)$ and $G_2 = (V_2, E_2)$. The join of G_1 and G_2 is written as $G_1 + G_2$ and is a graph with a vertex set consisting of $V_1 \cup V_2$. The edges are $E_1 \cup E_2$ along with an edge between every element of V_1 and V_2. For example, if our two graphs only have vertices and no edges, then the join of them is the complete bipartite graph of the edges V_1 and V_2. It also does not matter what order we join the two graphs in, $G_1 + G_2 = G_2 + G_1$.

The join is then a way to unify two graphs modeling diverse concepts. The method it uses, however, may not be the most illuminating. To illustrate this, consider the case where we have a graph that is modeling domain resolutions. The second graph is contains a single vertex representing 127.0.0.1. If we join the two graphs, then every vertex in the first graph gains a relationship to 127.0.0.1. This does not necessarily make sense to if there is an IP address in the first graph, because there is no real reason it should have a relationship with 127.0.0.1.

The second method we cover in this section is called the **Cartesian product**, or the **product**, or even the **cross product**. In this case, we have two graphs $G_1 = (V_1, E_1)$ and $G_2 = (V_2, E_2)$. The vertex set of the Cartesian product $G_1 \times G_2$ is the set $V_1 \times V_2$. That is, it consists of all the pairs (v_1, v_2) where v_1 is a vertex in V_1 and v_2 is a vertex in v_2. If (v_1, v_2) and (u_1, u_2) are vertices in $V_1 \times V_2$, then there is an edge between them if $v_1 = u_1$ and there is an edge in G_2 between v_2 and u_2 in G_2 or if $v_2 = u_2$ and there is an edge in G_1 between v_1 and u_1.

In the Cartesian product, we unite our graphs on the common vertices while maintaining the edges in each graph. So if we have a graph modeling domain resolutions and one modeling IP addresses accessed in the network, we unite the two graphs on the common IP addresses between the two and create relationships between the IP addresses visited and the resolution of the IP addresses.

5.12.4 THE INTERSECTION GRAPH

The intersection method produces a graph that starts with a collection of sets of data. The method creates a graph called an **intersection graph**. Consider a collection of sets $\mathcal{F}$ where each element of $\mathcal{F}$ is a set S_i. For example, each set in the collection could contain all of the IP addresses a user accessed in a day, or each set contains the domain names a piece of malware accessed, or the registry keys that malware altered, or the set could contain the people that attended a particular event.

The vertices of the intersection graph are the sets in $\mathcal{F}$. There is an edge between two sets if $S_i \cap S_j \neq \emptyset$. An interesting theorem in graph theory says that every graph can be written as an intersection graph. This is actually easy to do. Let $G = (V, E)$ be our graph. For every vertex v_i in G we create a set S_i that contains the edges incident with v_i. If there is an element in common with S_i and S_j, then there is an edge incident with both v_i and v_j. In other words, we have created the same graph on the S_i vertices as we did on the vertices in V.

Example 5.12.1. A honeynet is a collection of honeypots, designed to attract attackers by the vulnerabilities built into the systems and the network. Each honeypot logs who attempts to attack it and what attack is used. An intersection graph is an easy way to see what commonalities across systems exist. For example, we could use as our sets the IP addresses that connect to each honeypot. After we create the intersection graph from this data we discover that one system has no IP addresses in common with any other system. The next step is to analyze why this is the case as well as to examine the IP addresses that other systems have in common.

5.12.5 MODIFYING EXISTING GRAPHS

We are given a set of users and the domains that they have all visited and used this to create a graph. At first analysis, we notice that every single one of our users has visited google.com and it seems to be skewing our results a bit. The logical thing to do is to remove the vertex that represents google.com. If we do that, then we have to remove all edges to which that vertex is incident to, because otherwise they make no sense. Remember, an edge is a model of a relationship and if one end of the relationship is removed, then there is no longer a relationship.

This technique is the same if we are working on any graph. We remove the vertex in question and all edges incident to that vertex.

On the other hand, if we remove an edge, we do not touch the vertices to which it is incident. We are just removing the relationship, not the entities that have the relationship.

Suppose our graph contains example.com and www.example.com. This is redundant, so we want to collapse our two vertices and replace it with one that represents both. To do this, we create a new vertex we will call example.com+www.example.com. This new vertex will be incident to every vertex both example.com and www.example.com were incident too.

Again, this works for graphs in general. This is collapsing an edge between two vertices to form a new vertex and remove the relationship. This occurs when analysis of the graph shows that the two vertices are essentially synonyms.

5.13 CONCLUSION

This chapter has covered the fundamentals of graph theory and given examples as to how it applies to cybersecurity. We can use these graphs to analyze data that has a relationship within it, such as DNS, BGP, malware authors, malware, and more. The underlying theme of the chapter is that it does not matter what data we are modeling, as the graph properties do not depend upon the data, rather the graph.

New methods to analyze cybersecurity data using graphs are cropping up in the literature often. For example, an algorithm to find DNS sinkholes uses graph analysis techniques on passive DNS. They have also been used to detect botnets as well as route injections. As we have said before, any data with relationships can be modeled with a graph. The important thing is knowing how to analyze the graph in order to discern patterns, important relationships, or anomalies.

CHAPTER

Game theory

6

Chess is a game played by two people where each player has a strategy. The goal of chess is to put the other player into a position where their king will be captured no matter what move they make. This would be a win for the player, or payoff. Poker is also a game where the players bet money based on their cards. The payoff is the money in the pot that the player gets for a winning hand. The loser in the game of poker also have a payoff, but in their case it is a negative, that is, a penalty. Players have strategies based on how good their cards are versus how good they believe the other player's cards are. The cards are hidden, meaning that the player does not have all of the information before making their bets. Further in poker, the players may be further unaware of an opponent's possible choices, might they be bluffing or willing to call our bluff in various situations? This is opposed to chess, where the player knows what moves the other player has made and where all of their pieces are.

Both chess and blackjack have basic themes in common. Players, strategies used by players, and payoffs. We can abstract these notions to create the mathematical field of **game theory**. Game theory is applied to many areas and is a useful way to quantify outcomes which may depend on the actions of multiple players or strategic agents. For example, we can model the network defender and network attacker interactions as a game. There are clearly two players, the defender and the attacker. The defender has strategies to protect the network while the attacker has strategies to breach the network. The payoff for the attacker are the digital assets (PI, data, secrets, intelligence, cryptographic currencies, bank account numbers, and more) they are able to steal from the defender, on the other hand the defender works to avoid the penalty of their loss.

Game theory is also applied to economics, as it is a way to model competition in a marketplace. In social contracts, game theory can be used to consider cooperation of people versus selfish actions. In general, it can be used to model competitions that are present in a variety of fields. This chapter covers basic games that occur often in game theory analysis as well as solutions of games. The solutions are the strategies used by the player to ensure the best result, or the highest payoff. A solution for a casino card game would involve a betting strategy to maximize expected winnings. Mathematicians have solved this problem for two person Texas Hold 'Em. They determined the best strategy to maximize the winnings in the game.

Cybersecurity and Applied Mathematics. http://dx.doi.org/10.1016/B978-0-12-804452-0.00006-3

6.1 THE PRISONER'S DILEMMA

We begin by examining a simple game and then use that to build the basics of game theory. The game is called the **prisoner's dilemma**. The game is a two-player game that models cooperation versus competition. It has applications in economics, politics, sociology, biology, psychology, and political science, to name a few fields. It has also been called the game theory equivalent of an eye for an eye.

The game scenario is that two people, Alice and Bob, have been accused of a serious criminal offense. Unfortunately for the police, they only have evidence of a lesser crime. Alice and Bob are put in separate rooms and given two options. They can stay silent or they can confess. If they both stay silent they will only be charged with the lesser offense. If one stays silent while the other confesses, the silent one will be charged with the serious offense while the other will go free. If both confess, they will both be convicted of the serious offense but the prosecutor will ask for lighter sentences.

Alice has two actions in the game. She can stay silent or she can confess. Bob has the same two actions and the result of the game depends on both Bob and Alice's actions. Bob's choices can influence his outcome and they can also influence Alice's. There are four possible outcomes in the game, we list them in Table 6.1. The S stands for silence while the C stands for confess. Alice and Bob have to chose their strategy based on what they think the other player will do.

The payoff for each player is the number of years they are sentenced to in jail. Clearly, the player wants to spend as little time in jail as possible. For example, let us suppose that if both stay silent, they get 2 years each. If one confesses and the other stays silent, the confessor gets zero years while the one that stays silent gets 20 years. Finally, if both confess, they will get 10 years each. There are four possible payoffs for the game, we list them in Table 6.2. This matrix of payoffs is called the **payoff matrix** for the game.

If Alice stays silent, the best she can do is 2 years in jail while the worst is 20 years. If she confesses, the best she can do is no years and the worst she can do is 10 years. The dilemma for Alice is that while confessing holds the chance for the best outcome for her as an individual, staying silent would give the best result for them as

Table 6.1 The Outcomes of the Prisoner's Dilemma

	Alice	Bob	Alice	Bob
Alice	S		C	
Bob		S		S
Alice	S		C	
Bob		C		C

Table 6.2 The Payoff Matrix for the Prisoner's Dilemma

	Alice	**Bob**	**Alice**	**Bob**
Alice	2		0	
Bob		2		20
Alice	20		10	
Bob		0		10

a pair. Does she choose what is best for herself or does she choose what is best for the group.

Rather than using the terms *silent* and *confess*, this game sometimes uses the terms *cooperate* and *defect* to make it clear that it is about the group versus the individual. The player that stays silent cooperates with the group for the greater good of the group. The player that confesses defects from the common good for the betterment of the individual.

6.2 THE MATHEMATICAL DEFINITION OF A GAME

In the previous section we discussed the prisoner's dilemma with two players. There is actually no rule that requires that game to be played with only two players, that is just the simplest version of the game. In this section we will discuss the mathematical definition of a game with n players. We will assume that $n \geq 2$ as we are not discussing single player games. Game theory is a study about competition. This means that there must be at least two players for the competition to exist.

6.2.1 STRATEGIES, PAYOFFS AND NORMAL FORM

The set of players is $\{1, 2, \ldots, n\}$. Each player i has a set of strategies (or moves) S_i. It is entirely possible that $S_i \cap S_j = \emptyset$. The set of moves may nor may not be disjoint. From the prisoner's dilemma game, we know that $S_1 = S_2$. If we consider the game between the network defender and the network attacker, then we do not expect them to be using the same strategies.

Each move has a potential payoff. So if there are m strategies for a player, there are m payoffs. The set of payoffs for each player's strategies is U_i. So each strategy $s \in S_i$ has a payoff $u \in U_i$. Each payoff is a real number and is associated with a strategy, so rather than writing this as a separate set we can write it as a function. It is not just the player's strategy that can affect the payoff, but the other players in the games can affect the payoff. We saw this in the prisoner's dilemma. The payoff function is a function $u_i : S_1 \times S_2 \times \cdots S_m \to \mathbb{R}$.

The easiest case to visualize is the two player game. $u_1 : S_1 \times S_2 \rightarrow \mathbb{R}$ and $u_2 : S_1 \times S_2 \rightarrow \mathbb{R}$ are the payoff functions. We can list all of the possible payoff outcomes in matrix form. We will let A_1 be the payoff matrix for player 1. Each element of the matrix $a_{i,j}$ is the value of the function u_1 at (s_i, s_j). In short, $a_{i,j} = u_1(s_i, s_j)$. In a similar fashion we can create the matrix A_2 for player 2. In Table 6.2 we combined the two payoff matrices into one table for easier viewing.

Every time a player executes a strategy $s_i \in S$, the opponent chooses a strategy of their own. If we do not know what that move is, or if we just want to denote that the second player made a move, we denote the second player's action by s_{-i}. The payoff for a player's action s_i and the opponent's move s_{-i} is $u_1(S_i, s_{-i})$.

A game is called **symmetric** if $A_1^t = A_2$. The matrix equality implies that the payoffs depend only on the strategies chosen, not the player. More precisely, the same strategies apply to each player. The payoffs stay the same when we swap strategies between the players. The prisoner's dilemma is a symmetric game but most games are not.

If a player executes n strategies in a game, we can consider that an element of S^n. Depending on the game, a player can choose the same strategy repeatedly creating a vector $(s_i, s_i, \ldots, s_i)$ or can vary the strategy creating a vector with different elements. The vector is called the **strategy vector**.

A **strategy profile** is a vector that represents the set of all strategies at a round. It must contain one strategy for each player. For a round of the prisoner's dilemma, the strategy profile could be one of the four vectors $(C, C), (C, D), (D, C), (D, D)$. The payoff function is actually a function on the strategy profiles in the game to the real numbers.

We can also examine the individual moves by a player. This is a vector in S_i^m and can be written as $s = (s_p, s_q, \ldots, s_t)$. It is written like this because a player does not have to execute the moves in the order we list them in S_i. They can execute them in any order they wish.

6.2.2 NORMAL FORM

The **normal form** of a game is a matrix that describes the payoffs for the combinations of strategies. It is not a graphical representation, but rather a method to combine the strategies and payoffs in one location. One way to think of it is as a matrix of strategy profiles. In the previous section we discussed the strategy profiles for the prisoner's dilemma, which are $(C, C), (C, D), (D, C), (D, D)$. We can combine that with the possible strategies and create a matrix. This is described in Table 6.3.

The matrix is normally populated with the values of the payoff function, which allows us to make assumption about the game and actually displays rather quickly whether or not the game is symmetric. The normal form with payoffs for the Prisoner's Dilemma is in Table 6.4.

Table 6.3 Normal Form for the Prisoner's Dilemma

	Cooperate	**Defect**
Cooperate	(C, C)	(C, D)
Defect	(D, C)	(D, D)

Table 6.4 Normal Form for the Prisoner's Dilemma

	Cooperate	**Defect**
Cooperate	(2,2)	(20,0)
Defect	(0,20)	(10,10)

6.2.3 EXTENSIVE FORM

The normal form is a method for describing a game. Another method for describing a game is called the **extensive form**. It is a visual representation of the players and the moves they could choose. It also encapsulates what players know and the payoffs for strategies.

The visualization for the game is done as a graph with no cycles, or a tree. The vertices of the tree are the players and the edges are the actions that could be chosen by the players. It is also known as the **game tree**. Every vertex in the tree is a player and the edges that originate at a vertex are the moves. The tree has a root, or a top vertex. This is generally associated with the first player.

An example of a game tree for the prisoner's dilemma is in Fig. 6.1. We let the D stand for Defect and the C for Cooperate.

Prisoner one is the first player to move and he has two strategies. Once he chooses, then Prisoner two has two choices she can make. In total, there are four possible strategies for Prisoner two, each based on the choice Prisoner one makes.

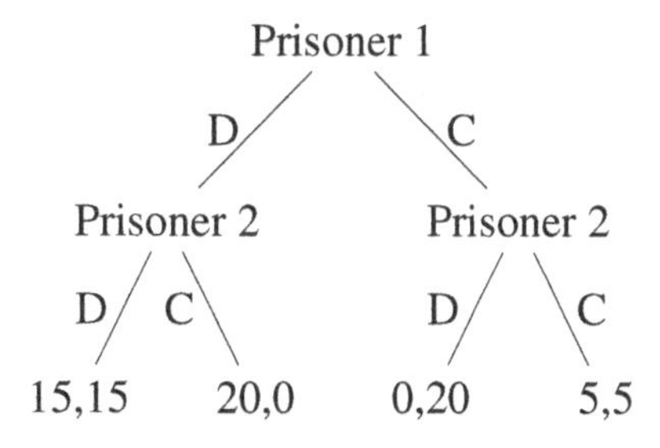

FIG. 6.1

Extensive form for the Prisoner's dilemma.

This encapsulates the same information in the normal form, but it is a more visual representation. We can easily trace the various outcomes given the possible choices for each player.

The extensive form has limitations. Let us look at the first level of the game tree for tic-tac-toe. There are nine possible choices for the first player to choose making the degree of the root nine. The second level has eight possible choices for each move player one made in the first round. Thus, we have 81 possible vertices in the first two levels of the game tree. As we play each round, up to nine possible rounds, that number grows. Level three has seven choices for each of the 81 vertices in the second level, or 567 vertices. This quickly makes the tree difficult to visualize.

6.3 SNOWDRIFT GAME

A game that models cooperative behavior is snowdrift. In prisoner's dilemma, cooperation means less punishment, not a gain for both sides. In snowdrift, cooperation is a slightly more attractive strategy plan than in the prisoner's dilemma game. But similar to the prisoner's dilemma, we do not know what the other player will do and are therefore partially informed.

In the game scenario, two people are traveling to their destinations by car. A large blizzard strikes the area, and creates a giant snowdrift. Our drivers are on either side of it that is blocking the road, preventing each from reaching their final destination. As luck would have it, each player has a snow shovel. Each player has a choice, they can do nothing or they can shovel snow. If both people do nothing, they cannot reach their destination until a spring thaw removes the snowdrift. If they cooperate and both shovel, they will eventually arrive at their destinations. If only one shovels, clearing the snow will take longer than if they work together. The lazy person who does nothing will gain his destination, but will have a less work involved.

The goal for each player is to get to their destination. Our two states are s_{shovel} and s_{sit}. Our payoff function u_i will be given by $u_i(s_{\text{sit}}, s_{\text{sit}}) = 0$, $u_i(s_{\text{sit}}, u_{\text{shovel}}) = 100$, $u_i(s_{\text{shovel}}, s_{\text{sit}}) = 300$ and $u_i(s_{\text{shovel}}, s_{\text{shovel}}) = 300$. The normal form for the game is in Table 6.5.

Clearly, the strategy that gives the best payoff for the player is to always shovel, or cooperate. This allows them to reach their destination whether or not the other player defects. If one player is sure of what the other player will do, the choice is easy. If you only have a belief of what they should do, then the best choice is to cooperate.

Table 6.5 Normal Form for Snowdrift

	Sit	**Shovel**
Sit	(0,0)	(100,300)
Shovel	(300,100)	(200,200)

Table 6.6 Normal Form for Information Sharing

	Cooperate	Defect
Cooperate	$(-Y, -Y)$	$(-(X+Y), Z)$
Defect	$(Z, -(X+Y)$	$(Z-X, Z-X)$

It is the only choice to guarantee the goal, that is, the reaching the destination. On the other hand, no matter what the other player chooses, cooperation means that every one in the game achieves their goal. The snowdrift game has been studied in comparison to the prisoner's dilemma and has been shown that it models cooperation more effectively.

Example 6.3.1. An ongoing issue in cybersecurity is information sharing. Suppose Bank Alpha has been attacked by a particularly vicious worm known as the Hodgins worm. Bank Alpha could share that information with Bank Beta so that they could be prepared for the worm, thus facilitating cooperation. Sharing information has a cost, but this cost is offset by receiving information from the other bank. If no information is shared, then the banks are going incur a cost due to the lack of information. In our example, Bank Beta would have to defend against the worm with no information. We assume that neither bank an take advantage of the other when they do not cooperate, based on a memorandum of understanding signed by both banks in regards to their cybersecurity.

Let the cost of information be X, the cost of having no information be Y and the value of information received be Z. So if we receive information and we spent nothing for it, we gained Z. If we spent a cost X to gain information valued at Z, then we have gained $Z-X$. If we spend X and gained nothing, then we spent $-(X+Y)$ and the cost of having no information is $-Y$. The normal form for cooperating (sharing information) and defecting (not sharing information) is in Table 6.6.

If would be best for the banks as a whole if they would cooperate, however, convincing them that it is in their best interest is hard.

6.4 STAG HUNT GAME

A variant of the prisoner's dilemma game is known as the Stag Hunt. This game is a very old one, first described by Rousseau in his book *A Discourse on Inequality* written in 1754. He said:

> *If it was a matter of hunting a deer, everyone well realized that he must remain faithful to his post; but if a hare happened to pass within reach of one of them, we cannot doubt that he would have gone off in pursuit of it without scruple...*

Rousseau was considering the case of a group hunting a deer, or stag, and one person who is pursuing the deer suddenly seeing a hare. Rousseau assumes that the

Table 6.7 Normal Form for Stag Hunt

	Stag	Hare
Stag	(300,300)	(0,100)
Hare	(100,0)	(100,100)

hunter will abandon the group hunt for the individual goal. We can take this story from Rousseau and model it as a game.

In our model of Rousseau's story, we begin with a group of hunters chasing a stag through the forest. If all of the hunters chase the stag, then the stag will be caught. If any one hunter decides not to hunt the stag and hunt a are instead, then the stag will not be caught but the hunter that defected from the group will catch the hare. The stag has more value than the hare, so if the hunters collaborate they stand to gain more as a team than if they defect.

In the case of two hunters, they can cooperate to achieve the shared goal of the stag or defect for the individual goal of the hare. If one hunter chooses the stag, he is assuming that the other will cooperate for the better payoff. If the hunter chooses the hare, then their payoff does not depend on the choices of the opposing player. This is in contrast with the prisoner's dilemma, where defecting (or confession) has a payoff that depends on the other player's choice.

The two states for the game are (S, H) for Stag and Hare. The payoff function can be defined as $u_i(S, S) = 300$, $u_i(S, H) = 0$, $u_i(H, S) = 100$, and $u_i(H, H) = 100$. The normal form for the Stag Hunt game is in Table 6.7. This encapsulates the idea that both hunters cooperating achieve the larger payoff, that is, the stag. If the defection occurs, then the defector receives the smaller payoff.

The strategy that guarantees a payoff no matter what the other player does is the hare. In other words, defection. Cooperation is dependent on the other players in the game. No payoff is available unless every player cooperates. This is in contrast to the snowdrift game, where cooperation ensures a payoff and defection does not.

6.5 ITERATIVE PRISONER'S DILEMMA

In the original premise for the prisoner's dilemma, the two players were given choices about their jail sentence. In this scenario, there is only one round of the game. The players choose once and are sent off to jail. We will consider a different scenario which still follows the basic tenets of the game. In this scenario, we will be able to play the same game over and over again, iterating the choices. The strategies are then a strategy vector and can be used to find a best solution.

Our game scenario considers two companies that compete in the same region. We have two Internet Service Providers (ISPs) and will call them ISP Omicron and ISP Theta. If both companies cooperate, that is, keep prices as they are, they will grow at the same rate. If ISP Omicron drops their rate (defects), then ISP Omicron will grow faster than ISP Theta. ISP Theta will remain stagnant or lose customers. If both drop prices, the increase in customers is spread equally between the two customers.

Unlike in the original version of the prisoner's dilemma, this can be played out repeatedly. In other words, a price war. The iterated version allows the player to remember the opponent's previous moves and to make moves based on them. This can affect the strategy that the player chooses. Each player makes a move and the other player considers that move before making a new move of their own. The goal of the iterative prisoner's dilemma is to maximize expectation.

One possible strategy is called the **tit-for-tat** strategy. The first move in this strategy is to cooperate. The next move is to the opposite of what the opponent did in their first move. So if the opponent cooperates on the first round, then the player defects on the second and if the opponent defects on the first round, then the player cooperates in the second.

In the price war, it can be considered a reactive response. ISP Omicron lowers prices, so ISP Theta lowers them in response. If ISP Omicron does not change prices, then ISP Theta follows that action. So the company is reacting to the previous round. If the prices are changed (defection) then in the next round, then ISP Theta punishes ISP Omicron for the defection. If ISP Omicron decides to cooperate, then ISP Theta forgives ISP Omicron for previous actions by positive reinforcement.

A variation of the tit-for-tat is called the **tit-for-tat with forgiveness**. It is generally the same as the tit-for-tat strategy, with a small change. If the other player defects, that is, changes prices, there is a small chance that the player will still cooperate rather than punish the player for his behavior. The small chance is generally around 1–5%. It is a slightly better strategy than the tit-for-tat strategy.

Another strategy is called the **tit-for-two-tat** strategy. From its name, we know that it is a variant of the **tit-for-tat** strategy. In this strategy, the player waits for the opponent to defect twice in a row before defecting. An alternative strategy is called the **two-tats-for-tit** where the opponent if punished for defecting. Every defection by the opponent is met with two defections by the player.

The political scientist Robert Axelrod studied the iterated prisoner's dilemma to determine the most effective solution. He created a computer tournament for the iterated game and had entrants create their own strategy to determine the best. His results indicated that the best strategies were not the greedy strategies. A greedy strategy is where the player defects more than his opponents, in our example, lowers their price the most. Instead, a cooperative, but not too cooperative strategy fared better. This strategy is called nice. Both retaliation and forgiveness were found to be part of a successful strategy. In short, the best strategy is a mixed of greedy and nice without skewing towards one or the other.

6.6 GAME SOLUTIONS

A player that plays a game would like to figure out which strategy is best. The players want to optimize the payoff they get from playing the game, assuming both players are rational with a common knowledge of the game. In other words, the player would like to figure out the best moves to guarantee the highest payoff. This can be a very complex problem, for example, in bidding for an auction. The player knows their strategy, but the other bidder may have different strategy which is unknown to the first player. Determining the best bidding strategy when the other strategies in the game are hidden from the player is very difficult.

On the other hand, a solution for the prisoner's dilemma is quite easy. If the player always defects, then they gain the best result. For snowdrift, the best solution is to always cooperate. The goal of game theory is not only to specify the games formally, but to determine that strategy that promises the optimum result.

6.6.1 COOPERATIVE AND NON-COOPERATIVE GAMES

There are two kinds of game, the **cooperative** game and the **non-cooperative** game. In the cooperative game, there is a third party involved. This third party enforces agreements between the players of the game. A well known example of a cooperative game is the called the *Lady with the bags*. A lady flies into a city with a lot of luggage. She has so much luggage that she requires two skycaps to carry the bags for her. The lady has a fixed amount to tip the skycaps and she has determined that each skycap's tip is determined by the number of bags each carries. It it is up to the skycaps to determine how to carry the bags to maximize their tip. The third party in this case is the lady with the bags while the players are the two skycaps.

Another example of a cooperative game is Dungeons and Dragons. The Dungeon Master is the third party that enforces the agreements between the players. Players make moves in the game predicated upon the enforcement of the rules in the game by the Dungeon Master.

The non-cooperative game is where the players make moves independently. The prisoner's dilemma, Stag Hunt, Blackjack, and Chess are all non-cooperative games. In fact, most recreational games are non-cooperative games. The game solution for a cooperative game is a strategy vector that defines the moves in the game that will optimize payoff. In the case of the prisoner's dilemma, we know that the solution of the game is to always defect. For a game like chess, the set of moves to gain the payoff is a very difficult problem. The goal of game theory is to find the optimum set of moves for a game, that is, to solve the game. The game solution is also called the equilibrium.

6.6.2 ZERO SUM GAME

The two person **zero-sum** game is when the gain of one player is equaled by the loss of the other. Chess, checkers, go, and tic-tac-toe are all zero sum games This is because there is a clear winner, and therefore a clear loser in these games. In the case of a tie, no one gains and no one loses. We can also consider a zero-sum game to be one where the players are in direct conflict. They have the same goal and one gains the goal while the other loses it.

If the players are not always competing for the same goal, then the game is not zero-sum. For example, if both players choose cooperation in the prisoner's dilemma, they both gain the same goal. So the prisoner's dilemma is not a zero-sum game because there is no direct competition with one player winning and one player losing.

A simple payoff matrix for a zero-sum game is in Table 6.8.

A game where there is a clear winner and a clear loser is a zero sum game. We can define the payoff for a win as a +1 and the payoff for a loss as −1. Thus the gain of the winner is equal to the loss of a loser.

6.6.3 DOMINANT STRATEGY

In Section 6.1 we said that in the prisoner's dilemma, the player should always confess, that is, defect from the other player. If player 1 does this, then $u_1(D, s_{-i}) \geq u_1(C, s_{-i})$. This is called a **dominant** strategy. This means that the payoff for this strategy is better than the payoff for any other strategy the player may choose. It is also a solution for the game.

To put it precisely, a strategy $s \in S$ for a player i is a dominant strategy if $u_i(s_i, s_{-i}) \geq u_i(s', s_{-i})$ for all $s' \in S$. As we saw in the previous discussion, the prisoner's dilemma has a dominant strategy. A game that does not have a dominant strategy is rock-paper-scissors, also known as Rochambeau. There is no move that will give the best payoff in this game, or in simpler terms, guarantee a win. No matter what move you choose, the other player can win and vice versa.

6.6.4 NASH EQUILIBRIUM

The dominant strategy is a game solution. It is also very rare that each player in a game has a dominant strategy. A less stringent requirement is the **Nash equilibrium**. This concept was created by John Nash, the Nobel prize winning mathematician who was the subject of the movie A Beautiful Mind.

The pure strategy Nash equilibrium is a set of strategic options selected by all players, such that no player can improve rewards by changing their strategy option while all other players' selection remain constant. In a way, we can think of this as a local optimization concept, and game player matrices can also give rise to multiple Nash equilibria.

Table 6.8 The Payoff Matrix for the Zero-Sum Game

		Player 1		**Player 1**	
			Player 1		**Player 2**
Player 1		x		y	
	Player 1		$-x$		$-y$
Player 2		$-w$		$-z$	
	Player 2		w		z

If we have our two players Alice and Bob from the prisoner's dilemma, then the strategy Alice chooses is predicated on Bob being rational and making the same best choice as Alice. If Bob changes his strategy, then Alice's payoff will be no worse than what she started with. If Alice chooses to defect, then Bob would make the best choice as well and also defect. If Bob changes his strategy to cooperate, then Alice's payoff will be no worse than what she started with. Therefore, the Nash equilibria is (D, D).

For the Stag Hunt game, the profiles that guarantee the best payoff are (S, S) and (H, H). That is, both players cooperate or both players defect.

6.6.5 MIXED STRATEGY NASH EQUILIBRIUM

A **mixed strategy** combines the pure strategy Nash equilibrium with probabilities. In short, a player has a pure strategies available, but chooses them based on probabilities. Suppose in the iterated prisoner's dilemma, the player chooses to defect 90% of the time and will cooperate 10% of the time. This is an example of taking a mixture of pure strategy by use of probabilities. The mixed strategy actually works best on games that play for more than one round thereby making the expected outcomes more important as time increases, and this is due to the powerful limit laws of probability such as the central limit theorem.

Nash actually showed that if multiple pure strategies existed for a game, then there is an equilibrium based on the mixed strategy. So if the game has a solution without randomness, then there is a solution if we combine the strategies with randomness.

Example 6.6.1. Let us reconsider the Stag Hunt game. We have two hunters, Neon and Boron. Neon decides he will hunt the stag with a probability of p and the hare with a probability of $1-p$. Boron's payoff for hunting the stag is $300p + 0(1-p) = 300p$. To put it simply, if Neon decides to hunt the stag 50% of the time, then Boron will get a payoff 50% of the time he hunts the stag.

Boron has decided he is indifferent to which animal he hunts. This means that to him, the payoff for hunting stag or hare is exactly the same. We can solve the equation $300p = 100$ and get $p = \frac{1}{3}$. This means that if Neon hunts the stag $\frac{1}{3}$ of the time, then Boron will get the same payoff no matter which animal he hunts.

The exact definition for a mixed strategy is a probability distribution over the set of available strategies. If we have a strategy set of $S = \{s_1, s_2, \ldots, s_n\}$ then we expect a set of probabilities $P = \{p_1, p_2, \ldots, p_n\}$.

6.7 PARTIALLY INFORMED GAMES

A **partially informed** game is one where the players are missing key details. For example, in five-card study, the opponent sees all of the players cards except for the hole card. This card is kept hidden from the opponent and is used in the player's strategies. These games are also called **imperfect information** games.

A partially informed game is where the players know what possible strategies are, but they do not necessarily know the action. A network attacker may know the possible strategies a network defender could choose, but may not know the action the defender chooses. The attacker would just know if the attack succeeded or failed.

Example 6.7.1. A common example to illustrate the partially informed game is called the **Battle of the Sexes**. A couple, Dave and Carol, have agreed to a date but cannot recall the event they have planned on attending. The two events are a movie and the opera. One prefers the movie over the opera and vice versa. They begin the evening at different locations and would prefer to attend the same event. Assuming they cannot communicate to coordinate, the normal form for this model is in Table 6.9.

The model has perfect information for each player. Suppose that Carol wishes to avoid Dave, but Dave is not aware of that. In Dave's view, Carol wishes to meet and he will choose accordingly. Carol is aware of Dave's view of the date and can also choose accordingly. With that in mind, the normal form for Dave's evening is in Table 6.9. The normal form for Carol's evening is in Table 6.10.

Example 6.7.2. In a busy real estate market, a common tactic is the sealed bid auction. Rather than dealing with a list of offers and counter offers, the seller of the house tells the interested parties to submit a single bid in a single envelope. The seller opens the bids and the person with the highest bid wins.

We will consider the two player case. Two people submit sealed bids for the same house. If the bids are equal, then the decision is made by coin flip. A player knows the value they put on the house and the amount of their bid. They do not know the value the other player puts on the house or their bid. We will let v_i be the value each player puts on the house and b_i be the bid that the player submits.

The payoff for the player if the bid is accepted is the value $v_i - b_i$. The player has valued the house at v_i and offered b_i, meaning that the player gains $v_i - b_i$. If the

Table 6.9 Normal Form for Battle of the Sexes

	Movie	Opera
Movie	(3,2)	(0,0)
Opera	(0,0)	(2,3)

Table 6.10 Normal Form for Battle of the Sexes for Carol

	Movie	Opera
Movie	(3,0)	(0,2)
Opera	(0,2)	(3,0)

bid fails, the payoff function is 0. If the bids are equal, then the value that the player gains is $\frac{1}{2}$ of the payoff they expect if the bid was accepted. The reason for this is that the probability of winning the bid is $\frac{1}{2}$, based on the coin flip. So the payoff is the probability of winning the house multiplied by the actual payoff the player would gain if they won the house outright. Eq. (6.1) quantifies this payoff function:

$$u_i(b_i, b_{-i}) = \begin{cases} v_i - b_i & b_i > b_{-i} \\ \frac{1}{2}(v_i - b_i) & b_i = b_{-i} \\ 0 & b_i < b_{-i} \end{cases} \tag{6.1}$$

In Example 6.7.2 we saw that probability of the actions affected the payoff. This is common in the partially informed games. We know that there are possible strategies for the player to choose and what the probability is that a player will choose them.

6.8 LEADER-FOLLOWER GAME

In the **leader-follower** game, the leader is the player in the game that acts first. She chooses her strategy and the follower must react to that strategy in their next move. The leader can choose her strategy to increase her payoff. This increase in payoff is called the **leader's advantage**.

In our analysis of this game, we assume that the follower's strategy is rational. If our leader chooses strategy s_1, then our follower chooses the rational strategy $R_2(s_1)$. This is the response for our follower to the leader's strategy s_1. The payoff for the leader can be given as $u_1(s_1, R_2(s_1))$. The leader's payoff is then dependent only on her choice of strategy.

The leader has private information that the follower lacks. Thus, the follower's only information comes from the strategy that the leader chooses. The leader's goal is to maximize her payoff function $u_1(s_1, R_2(s_1))$ utilizing her private information and the knowledge that the follower will choose the rational reaction to her strategy.

A famous model of the leader-follower game is from Spence in 1974 in his model of the job market. The leader in this game is the worker and the follower is the firm or firms that are looking to hire. The leader knows her level of productivity and must choose how much education she would like to have. She can choose how much college education or job related certificates she would like to have. The follower, or firm, only knows her education level. They must make their job offer based on their belief in education level. In other words, they use the level of education to determine how much productivity the worker will have.

6.8.1 STACKELBERG GAME

A Stackleberg game is a two player game that models an economic strategy. The game scenario begins with a duopoly in an industry. Two firms are the dominant suppliers for a commodity in a region. One is the leader and the other is the follower.

For example, one firm could be a local firm with deep knowledge of the region and the other firm, the follower, is a company from outside the region attempting to make inroads. We will call the leader Firm 1 and the follower Firm 2 and they both create the same widget known as the doodad.

Firm 1 will determine the quantity of doodads it will create, then Firm 2 will use that number to determine the number of doodads it will create. The goal, or Nash equilibrium, is to figure out the number of doodads to create that will maximize the profit.

The profit depends on the price and the price can affect the number of the products sold. The higher the price, the fewer profits sold. The lower the price, the more products sold. We can plot this on a straight line as $P = a - bq$ where P is the price, a is the price at which no units are sold, b is the slope of the line and q is the quantity. For ease of computation, we will assume the slope is 1.

If q_1 is the number of doodads Firm 1 creates and q_2 is the number of doodads Firm 2 creates, then our function becomes $P = a - b(q_1 + q_2)$. Firm 1, since it has knowledge Firm 2 does not, can figure out how many doodads Firm 2 will want to create to maximize its profit. Firm 1 can then use this number to figure out how many doodads it should create to maximize its profit.

Firm 2's potential revenue is the number of units created multiplied by the price. So if we assume it sells everything it creates, their revenue is in Eq. (6.2):

$$R = Pq_2 = aq_2 - q_1q_2 - q_2^2 \tag{6.2}$$

Profit is given as revenue minus cost. If we assume constant cost c_2, then to create the cost to create q_2 units is c_2q_2. In order to maximize the profit, we take the derivative of the function $P = R - C$ with respect to the quantity and set it to zero. Then we can solve for q_2 which will give us the number of doodads Firm 2 must create given that Firm 1 has created q_1 elements in order for it to maximize profit. Eq. (6.3) is the solution:

$$q_2 = \frac{(a - q_1 - c)}{2} \tag{6.3}$$

Firm 1 can make this calculation as well. They can take this value for q_1 and use it to maximize their profits. The number of products they will need to compute to maximize their profits is given by $q_1 = \frac{a-c}{2}$. Then the best response for Firm 2 is $q_2 = \frac{a-c}{4}$.

6.8.2 COLONEL BLOTTO

The game **Colonel Blotto** is a leader follower game that considers strategic moves as part of the scenario. It also considers allocating resources. The premise is that Colonel Blotto and Colonel Zappo have 10 men each to allocated to 5 fronts in a war. If Colonel Blotto has allocated more men to a front than Colonel Zappo, then Colonel Blotto wins the battle on that front. If they have allocated an equal amount of men, no

Table 6.11 Colonel Blotto Versus Colonel Zappo

	Front 1	Front 2	Front 3	Front 4	Front 5
Blotto	3	2	1	2	2
Zappo	1	3	2	3	1

one wins. If Colonel Zappo allocates more than Colonel Blotto, then Colonel Zappo wins. The payoff for each colonel is the number of fronts that colonel has won minus the number of fronts that colonel has lost.

A simple example of Colonel Blotto is in Table 6.11.

In this example, Colonel Zappo has a payoff of 3 where Colonel Blotto has a payoff of two, meaning Colonel Zappo has won this round. If Blotto knows how Zappo allocated his men, then Blotto could easily reallocate his men to win the next round.

This is an example of a two player zero-sum game, when Blotto wins one, Zappo loses one, and vice versa. However, the game does not have an optimal pure strategy for winning the game. There are many mixed strategy equilibria for the game though. One possible mixed strategy equilibria assumes that Blotto and Zappo allocate their troops across the fronts by proportion. The proportion of troupes stays the same, even if they are swapped between different fronts. In other words, we allocate them as (4,3,2,1,0) across the five fronts, where Blotto could allocate 4 in Front 1 or in Front 3. Zappo also uses the same proportions.

There is no requirement that the Colonel Blotto game is played on only 5 fronts. We used 5 to illustrate the game. The game can be played on any number of fronts, as long as there are more than 2 fronts. If there are only 2 fronts, then the game always ends in a draw. It has been proved that the probability distribution for the mixed strategy of Colonel Blotto on n fronts is a uniform distribution on $[0, \frac{2}{n}]$.

If we change our scenario slightly, the Colonel Blotto game applies to cybersecurity. We have two players, the bad guy who uses a DDOS to attack the good guy. The good guy has a number of geographically diverse facilities but only a limited amount of mitigations available to counter a DDOS. The bad guy wants to take down the maximum number of facilities that the good guy owns, while the good guy must determine the best allocation of his mitigation resources to repel the DDOS and maintain an internet presence.

6.9 SIGNALING GAMES

A signal is something that is designed to convey information. The diploma on your wall is a signal that conveys to the onlooker that you have been educated. This does not actually mean you are educated, merely that you are signaling to the world that you are. This implies that a signal does not have to be true, it can also be false. Suppose a celebrity has endorsed a particular product. That endorsement can signal that it is a good product, however, it can also signal that the company that makes the product paid the celebrity to endorse it. In this case, the signal can be misleading. Another example is a certificate you could have on your wall. The certificate may say MCSE, but without explaining what MCSE stands for, the signal could be very confusing. The signal suggests but doesn't guarantee the attributes of the sender, so we say that the attributes of the sender are incomplete information.

The signals could have gains or loses depending on how they are read and what they mean. If you signal a diploma it could land you a job. If the diploma is fake, your deceptive signal can land you a job outside of your reach but that may come with the danger of losing the job once you are found out. If your diploma is real, then you can prove it if questioned thus leading to additional job security. The same holds true for the celebrity endorsement. The company that uses that signal could gain in sales based on it. If the sender of the signal, the celebrity, loses credibility, then the company could lose sales based on the signal. For example, if the celebrity has a scandal that damages their reputation in the press, then any company that uses that celebrity as a spokesperson is in danger of losing sales.

A move in a perfect information game is a signal, but it is a truthful signal with perfect information. When a player moves a piece in chess, she is signaling to the other player her intentions. It is up to the other player to read the signal and determine the intentions.

A **signaling** game is an imperfect information game consisting of two players. Player one is the **sender**, that is, the person that sends the signals. Player two is the **receiver**. The receiver receives the signals that the sender sends and determines his action based on the information.

An element of a set T is chosen at random and assigned to the sender. The element $t \in T$ is known as the **type** of S. In general, it is assumed that nature picks the type of the sender and neither the sender nor the receiver can affect that selection. The sender chooses an element $m \in M$ and sends it to the receiver. The set M is a set of signals related in some fashion to the set T. The receiver takes the value of m and chooses an action from a set A to execute. The sets M and A do not depend on the choice of type nature makes in the beginning of the game.

To put this in terms of sets and functions, we have a function $m : T \to M$ for the sender and a function $a : M \to A$ for the receiver. The sender defines the function $m : T \to M$ and the receiver has no knowledge of what that is. It could be a truthful function, or it could be a misleading function. The sender sends $m(t)$ to the receiver and the receiver determines $a(m(t))$ for the action that the receiver takes. The game is the composition of the two functions.

Example 6.9.1. A common example used to illustrate the signaling game is the Beer-Quiche game. Player one has one of two types, either weak or strong. We will designate them as t_{strong} and t_{weak}. There is a probability associated each type, where p is the probability that player one is strong and $1 - p$ is the probability that he is

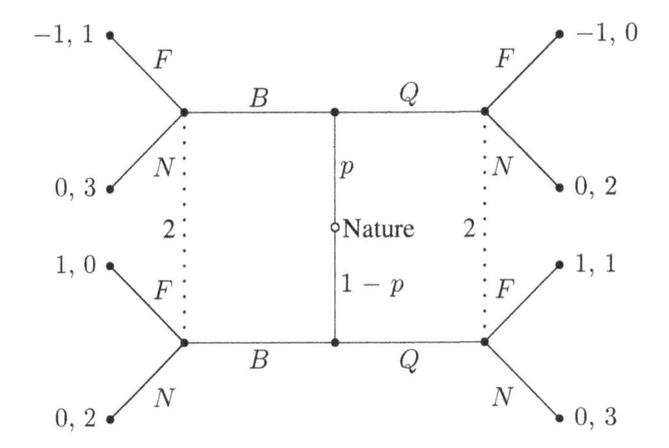

FIG. 6.2

The beer-quiche signaling game.

weak. Player two is a bully and wants to fight player one, but only if player one is a weak. He has no desire to fight someone just as strong as he is, he wants a win. On the other hand, player one has no desire for a fight. He uses his breakfast to signal his type to player two. If he has a beer for breakfast, then he is strong but if he has a quiche for breakfast, then he is weak.

The Beer-Quiche game is summarized in Fig. 6.2. *B* denotes Beer, *Q* denotes Quiche, *F* is fight and *N* is not fight.

Signaling games are useful in cybersecurity as most users, whether they are attackers or not, pretend to be cooperative users in order to gain access necessary for attacks. Once they gain access, they may launch an attack. This scenario played by the attacker (sender) and defender (receiver) can be modeled by signaling games. The conflicts may be represented by zero-sum quantities or costly payoffs for trusting someone who ultimately turns out to be an attacker.

CHAPTER

Visualizing cybersecurity data 7

"A picture is worth a thousand words." We know that an image can convey a complex thought more effectively than words. Similarly, if we have collected a set of data, a picture used to illustrate it can be more effective than just talking about the data.

For example, there are 65,535 ports available on a networked computer through which traffic can flow. We want to know the activity on a system over a period of time to look for anomalies, one method to do this is to make a table of all of the ports that were active along with the activity times. If we assume that there are 50 lines on a page, then that is over 1300 pages of data if we list all of the ports available on the system. Finding anomalies this way is a tedious job because there is no way to easily see an overview of time versus ports accessed by using a table. On the other hand, if we made a plot on the xy plane where the x-axis is time and the y-axis is the ports, then we can see anomalies or patterns.

Visualization is a model for the eyes. It has assumptions underlying it that turn the human eye into a decision maker. Because of this it is important to consider the purpose for a visualization when designing it. If a decision is to be made based on the image, then it should illuminate the data to assist in the decision. If the visualization obscures information or confuses the issue, then the visualization should be modified. If the creator of the visualization is the only person that can understand the visualization, then it should be modified. It should model the data so that it is comprehensible with a minimum of explanation from the creator.

Unfortunately, cybersecurity has a history of not doing visualizations well. This chapter discusses deciding what to visualize and why visualizations are important. It also covers interesting and useful visualizations that are not common but pertain to cybersecurity data.

7.1 WHY VISUALIZE?

A statistical model has one of three purposes, prediction, information extraction or description of stochastic structures. The visualization model can be used to tell a story about the data, find anomalies in the data or summarize it. The summary is a way of seeing the big picture over the complete space of data.

Cybersecurity and Applied Mathematics. http://dx.doi.org/10.1016/B978-0-12-804452-0.00007-5

Using a visualization to tell a story is extremely useful when explaining cybersecurity to an uninformed person. Telling a person that there is a massive increase in traffic to a noncritical system that should not be occurring is one thing, presenting them with a graph that displays how the traffic patterns changed over time is another. Similarly, an image that displays the exhaustion of IPv4 space is more compelling than simply saying the exhaustion is imminent.

An anomaly is a data point that does not fit, so if we visualize the data, then anomalies should stand out. For the an example of network traffic, an anomaly in the data would be a sudden spike in traffic which we would expect to be readily apparent in a traffic visualization.

Network traffic has several features. It includes the number of bytes, the number of packets, the origin port, the destination port, the origin IP address, the destination IP address, the protocol, as well as the time. We can create network traffic visualization combines the bytes and the time, summarizing the interaction. We can create a visualization that includes more information. However, aggregation can obscure features, and we should be mindful of that when creating visualizations.

Visualizations can be used to drive further analysis of the data. Suppose we are looking for anomalies in the port data of a computer system. A scatterplot of ports versus time may highlight the fact that certain ports accessed over a time period in question, We may then find that these ports are associated with known vulnerability, therefore the system should be analyzed to determine if it has been compromised.

7.2 WHAT WE VISUALIZE

Before creating a visualization, we should consider our model. If we want it to highlight anomalies, then summarizing all of the data can obscure an anomaly in an active field. When creating a visual model, we need to consider the data in question. If we are creating a scatter plot on the xy plane, then we want to make sure that the plot tells a story, illuminates anomalies in the data or summarize it in a useful way. We need to ensure that the visualization does not confuse the viewer by obscuring interesting features of the data.

7.2.1 CONSIDERING THE EFFICACY OF A VISUALIZATION

Our data set is a collection of IP addresses and the ports on those IP addresses that were accessed over the previous 7 days. The first idea we have is to plot IP address versus port. That is, our plot is on the xy plane where the x-axis is the IP address and the y-axis is the ports. Fig. 7.1 is an example of this. The resulting plot is very busy.

There are several problems with this illustration. First if we notice an anomalous port, we have no sense of how often it was accessed during the time period in question. Second, if there are too many IP addresses, how do we figure out which IP address is the affected one? We can also consider the case of a false anomaly. Suppose we see in our port that most IP addresses have had port 1433 accessed. At

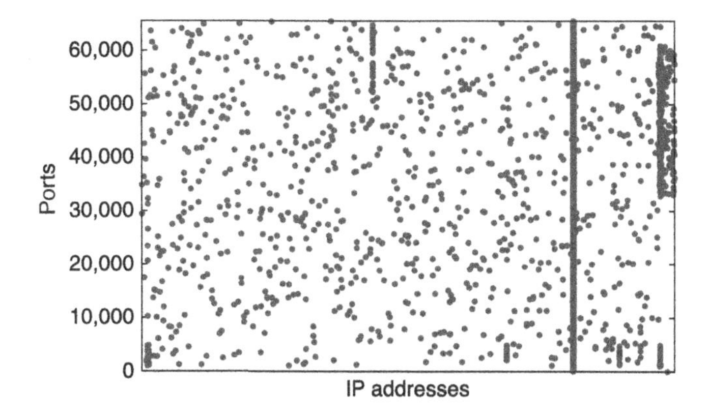

FIG. 7.1

IP addresses and active ports.

first glance, this could be construed as an attack. However, the Microsoft SQL server has been a known vulnerable port for many years. Traffic accessing this port is not an anomaly or directed attack, it is essentially the same as port scanning.

We can also see port scans in these results, assuming we restrict the time period in question to less than 7 days. Port scans are so common in the Internet that their presence in network data is not anomalous. The visualization has no clear pattern that would inform the next step of analysis. It also does not effectively tell a story, and the summary of the data is difficult to understand. It is a good example of what not to do and illustrates the first sentence of this chapter very well. We could say "visualizing this is a bad idea," however, the illustration in Fig. 7.1 makes the point for us.

We could turn this visualization into a three-dimensional visualization by adding a z-axis, which in this case would be time. The visualization is still very busy and it is still very hard to find patterns within the data. Trying to figure out exactly what pattern matches which system is difficult in the noisy data.

A better visualization is to pick critical IP addresses, for example, the web servers, the mail servers and the DNS servers. For these IP addresses we create a plot of time versus ports on the xy plane. By narrowing our view, we are examining activity on critical assets and looking for potential anomalies that affect those critical systems.

7.2.2 DATA COLLECTION AND VISUALIZATION

The data collected affects the visualization. If our data collection process changes, then the visualization can change in ways that affect the analysis. Suppose we are measuring network traffic with n sensors. After examining the network topology, it is determined that a portion of the network is not covered by the existing sensor network. A new sensor is added, and suddenly it appears that the amount of data traversing the network has jumped significantly. If our visualization plots time versus the amount of data traversing the network, then the new plot would look like Fig. 7.2.

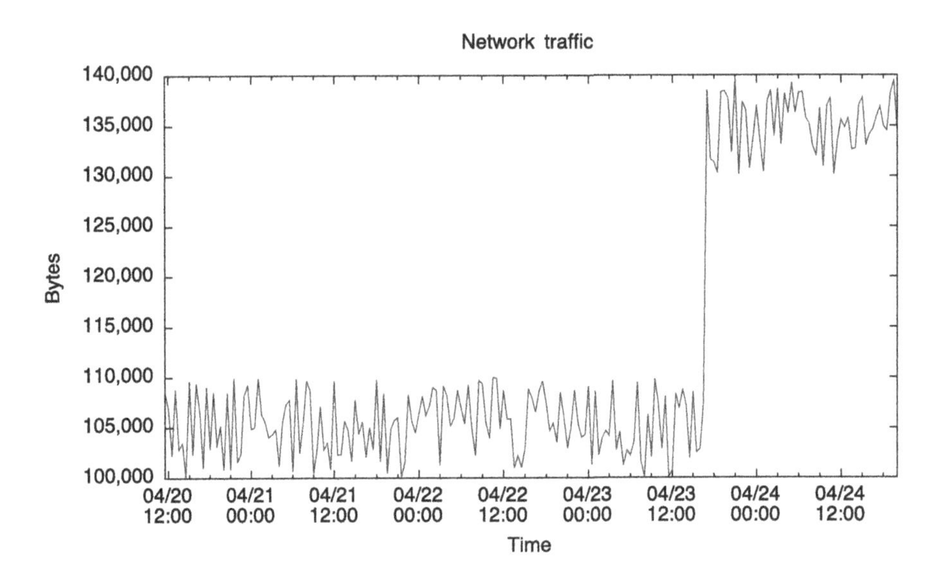

FIG. 7.2

Network traffic by time.

Without the knowledge of the additional sensor, then the visualization could be a cause for concern. Suddenly, the amount of traffic across our network has increased dramatically. If we are using this visualization to illustrate the amount of traffic crossing the network, then the sudden increase in traffic without context could skew decisions made with the visualization. Consistent data collection creates the best visualizations.

As we saw in Fig. 7.1, the more sources of data we aggregate, the less likely we are to know which source is the issue. By aggregating the IP addresses onto one plot, we lost the ability to determine which IP had the anomalous activity, even if we could find it.

If we reconsider the network traffic plot in Fig. 7.2, the aggregation of all of the traffic across all of the IP addresses hides the IP address that may be creating the most traffic. If the purpose of the visualization is to show the total traffic on a network, the aggregation of the IP addresses is a valid choice. If the purpose of the plot is to find the IP address with the largest amount of traffic, the aggregation will not work.

7.2.3 VISUALIZING MALWARE FEATURES

For another example, suppose we are attempting to visualize malware features. A piece of malware has a large number of features associated with it (eg, the size of the program, to the files it drops, the domains it attempts to access, the ports it used, the registry keys modified). How do we visualize a domain name or a dropped file? We could treat them as plain strings, but how would we create that visualization?

If we are trying to do a plot on the xy plane, we could count each feature. If we do that, we are comparing the size of the feature, not the feature itself. Suppose we have 100 samples of malware to visualize. If we plot the size of the malware and the number of files dropped, we have an extreme difference in scale. The size of malware is numbered in the thousands of bytes, while the number of files dropped tends to be an order of magnitude in difference. So our plot will hide the number of files dropped while attempting to display the sizes of the malware. One method to fix that issue is to use a log scale on the y-axis. The log scale prevents the smaller numbers from being hidden while still displaying the larger values.

Perhaps we should consider why we are creating this visualization. The size of a piece of malware does not necessarily correlate with the number of files dropped, so we need to examine what story we are trying to tell by displaying both features on the same plot. This does summarize two features of our set of malware, but why were those chosen and not other features? Is it possible that for one family of malware, the two features correlate, and that is our story? Or is it possible that these two features were chosen as the "most interesting" features? We should consider what makes those features interesting and why we display them.

The point of this discussion is to emphasize that when we start to create a visualization, we need to consider why the visualization is needed. Are we telling a story about two malware features, looking for anomalies in these features, or summarizing them?

7.2.4 EXISTENCE PLOTS

An **existence** plot is a low resolution time series used to study active ports on a host. It is similar to the scatterplot of time versus port mentioned earlier in the chapter, but displays more information. The plot summarizes either inbound or outbound activity for a single host in a two-dimensional plot and the summary is over all of the sources that the host communicated with. It is a coarse representation, but it illustrates all activity on that host.

In that coarse representation is a high level summary of the traffic to and from that host. The coarseness can obscure individual ports, but the patterns of traffic are apparent. Existence plots can be used to find hidden servers, that is, hosts acting as public servers that should not be. For example, malware can be used to subvert a system and turn it into a name server or a web server. An existence plot can be used to find that host. We create unidirectional plots, so each host is represented by two images.

We create this plot by considering a time interval τ and a port p. We find number of bytes that flowed through the port, either inbound or outbound, during the interval τ and denote it by $X_{p,\tau}$. $X_{p,\tau}$ is then assigned a color based on how much traffic flows through the port.

We divide the amount of traffic into three intervals. These are used to color the values plotted in each τ so that we can easily see where the high traffic areas are versus the lower traffic areas. We can divide this in one of two ways. We could measure the range of $X_{p,\tau}$ over all values of p and τ and divide it into the equal intervals of $(0, S_0)$, (S_0, S_1) and (S_1, ∞). This creates a color scheme that is unique to each plot, which can make it difficult to examine multiple plots. If two plots have the same amount of red, but the color means different range of byte counts, then it is tempting to correlate the two plots as showing an equal amount of data. We should be careful when we use this method to determine the color, particularly when creating multiple plots.

The alternative is that we have an idea of the magnitude of all traffic over the network. We can then fix the values S_0 and S_1 for all plots. This yields consistency of meaning across all plots.

Once we have our values S_0 and S_1, the colors are chosen based on the magnitude of the data. Eq. (7.1) is the equation used for color selection:

$$\text{color}(p, \tau) = \begin{cases} \text{none} & X_{p,\tau} = 0 \\ \text{blue} & 0 < X_{p,\tau} < S_0 \\ \text{green} & S_0 \leq X_{p,\tau} \leq S_1 \\ \text{red} & S_1 < X_{p,\tau} \end{cases} \tag{7.1}$$

Since we are grouping our time intervals, then the time series plot is not precise. Therefore we call it a low-resolution time series. The x-axis of our plot is time, broken into the intervals. Ports along the y-axis. We plot the ports in log scope, as the number of ports on a system is 65,536 and we expect most traffic to be concentrated on the lower range of ports. By using log scale, the lower range of ports is not squashed to the point of unreadability.

Fig. 7.3 is an example of an existence plot for inbound traffic on a host. In this plot, we see constant activity occurring inbound on port 25. We expect this host to be an SMTP server, and if it is not in our inventory as a one, it should be investigated further.

We are equally interested in activity and lack of activity on our hosts. Both of these can be illustrated by the existence plot. The lack of a straight line across the plot indicates that that port is not continuously active.

The value of τ also affects the plots. If we make it too small, then port cycles are less visible. If we make it too large, then port cycles over a shorter period of time are not visible.

Existence plots can be used to detect scans, hidden hosts, inactive ports, and port usage. It can be used to not only tell a story about the network traffic a system experiences, but to analyze the data to find anomalies.

7.2.5 COMBINING PLOTS

Suppose we have 10 hosts on our network and we would like to consider traffic patterns for all 10 of them. To accomplish this, we have measured the traffic to all 10 hosts over the previous week. The first thought is to create one plot that contains all

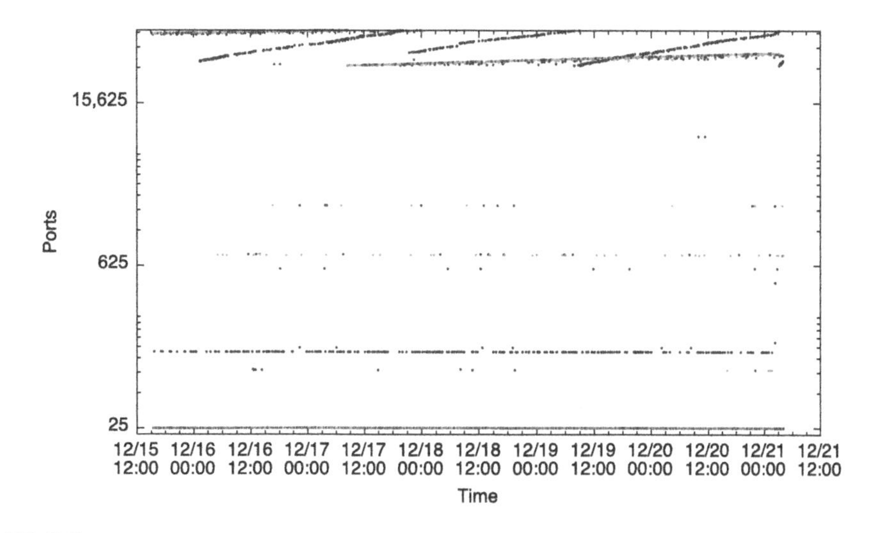

FIG. 7.3

An existence plot for InBound traffic.

of the data on it. This can be rather cluttered and cumbersome to read. If there is one host with a lot of traffic, for example, a web server and the rest of the hosts have very little traffic, then the scale of the plot will make it hard to see the patterns. Similarly, putting 10 distinct lines on one plot, even if they are different colors, turns the plot unreadable quickly.

As an alternative, we could make a series of strip plots. Fig. 7.4 illustrates that for five hosts. The eye can easily line up the plots in reference to time, particularly with the vertical reference lines included in the plot.

However, the y-axis has different scales for each individual plot in Fig. 7.4. The viewer must be aware of this, as otherwise the viewer can read inferences into the data that don't exist. The second and third plots have completely different ranges for the y-axis, one shows a low traffic pattern while the other shows a high traffic pattern. If we standardized the axis for this case, then the traffic in the third plot would not be visible in the plot. This is an example of how the standardization, which can make plots easier to understand, can also make the plots unclear.

It is important to note, however, that the time intervals on each plot must be the same. If two plots are for 1 week and the third plot is for 2 weeks, we should not line them up as in Fig. 7.4 without accounting for the difference in time intervals. Otherwise, the combination confuses the data rather than clarifying it.

7.3 VISUALIZING IP ADDRESSES

A visualization of the current usage of IPv4 addresses tells a story about the current allocation of IPv4 addresses. It is well known that almost all of IPv4 addresses are allocated, a visualization of that fact is a powerful tool.

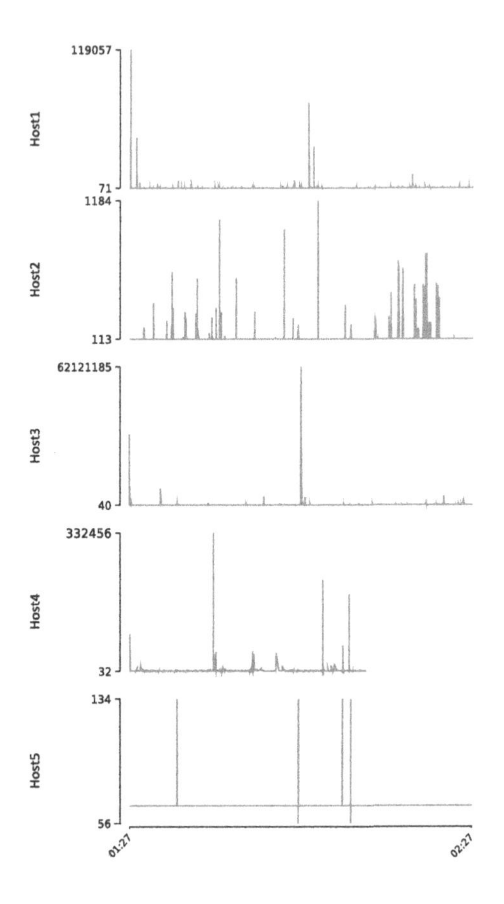

FIG. 7.4

Combined time series plots.

Visualizations of IPv4 addresses can be useful beyond displaying the current allocation of IPv4 addresses. If we are visualizing relative positions of IPv4 addresses, then this can be used to examine activity in IPv4 space. This can include A records in DNS and their relative positions, botnet activity and worm activity.

7.3.1 HILBERT CURVE

The **unit square** is the square with edge length of one and with corners $(0,0), (0,1), (1,0)$, and $(1,1)$ in the plane. A **space filling curve** is an injective function of an interval into the unit square, visiting each point in it exactly once. The curve has many applications, from mapping memory to database layouts.

The **Hilbert** curve is a space filling curve with an interesting property. Points on the curve that are close together in the unit square are close together on the line. The converse is not necessarily true. The curve is created by an iterative process, each iteration brings the curve closer to filling the unit square by increasing the amount of the unit cube that the curve fills. Fig. 7.5 is an example of the fourth iteration of the curve.

You can see how the curve bends as it winds its way through the unit cube. We have created a graph out of the unit curve because at each elbow, we put a vertex. The integers can then be mapped to this vertex. At the fourth iteration, there are 2^8 vertices. At the 8th, there are 2^{16}, at the 12th, there are 2^{24} and at the 16th, there are 2^{32} vertices. These numbers correspond to the number of IPv4 addresses in a /24, a /16, a /8 and the number of IPv4 addresses available.

The curve in Fig. 7.5 has been divided into four quadrants, each of which have the same number of vertices. Since there are 2^8 vertices in the curve, then there are 2^6 vertices in each quadrant. This maps to a /26 for each quadrant. Each of these quadrants can then be subdivided so that there are 16 (or 2^4) vertices in each quadrant. The nature of the Hilbert curve allows this subdivision to happen. From this notion we can infer that for the fourth iteration, the subnets of the /24 appear as either squares or rectangles. This idea is extended through the 16th iteration of the Hilbert curve. As that is the entire internet, then all possible CIDR blocks contained within the IPv4 address range would show as either a square or a rectangle.

We combine the Hilbert curve with a **heatmap** to create our visualization. A heatmap is a visualization of data where the color of the square is associated with the density of the population in the square. If the color is light, then the density of population in that square is light. The darker the color is, the denser the population. Fig. 7.6 is the Hilbert curve of the networks that are routed on the Internet as of November 1, 2015. We can see the density in the figure, the white areas are the ones that are not currently routed on the Internet while the black areas are fully routed.

FIG. 7.5

A fourth iteration Hilbert curve.

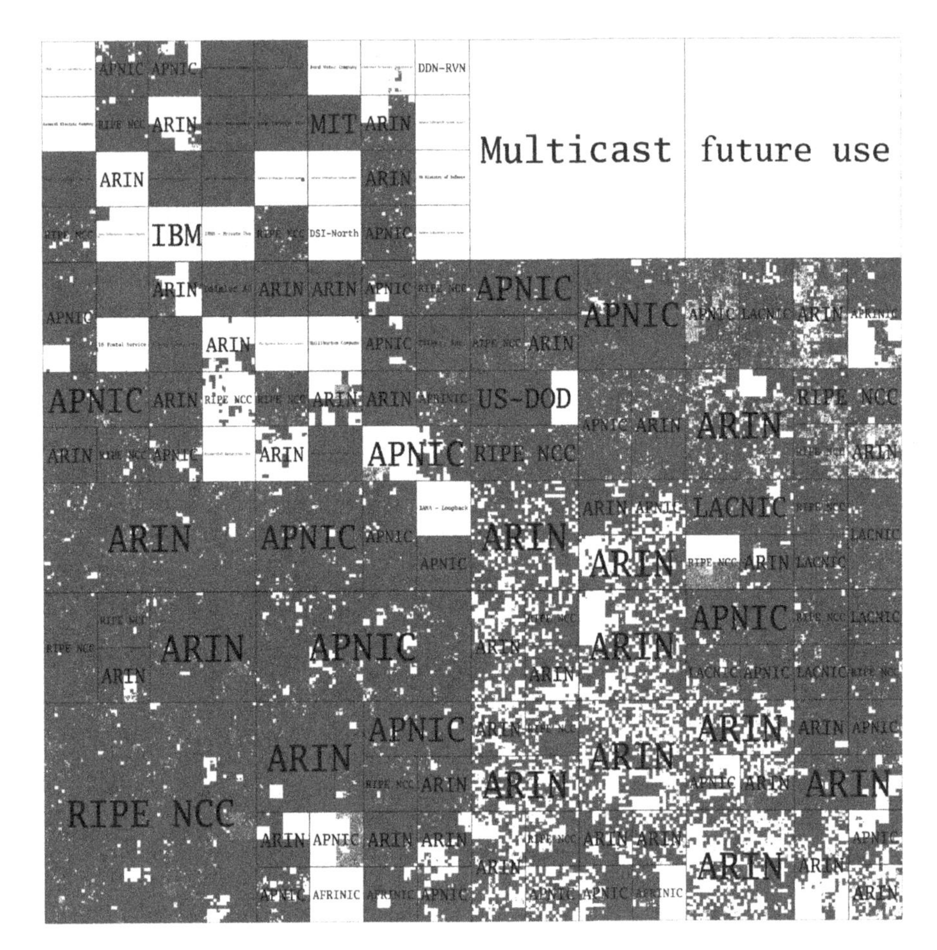

FIG. 7.6

Hilbert curve for IPv4 visualization.

Grey areas mean that subnets of the networks are routed, while there are some that are not. The darker the area, the more of the network is routed.

The overlay in Fig. 7.6 is very helpful. It segments the internet into related groups and enhances the Hilbert curve visualization. The overlay is based on the IANA IPv4 Address Space Registry. We can see clearly the different registrars and their allocations on the plot, along with the IPv4 address that is reserved.

Using the Hilbert curve as a visualization tool for IPv4 addresses was first proposed by Randall Munroe in his webcomic XKCD. His visualization at http://xkcd.com/195/ showed the groups of related addresses, taking advantage of the Hilbert curve and its inherent property. IPv4 addresses that are close together on the map tend to be close together on the interval. This is a great method for telling a story about the IPv4 addresses, but it is not very useful for analyzing the addresses. We can see clusters, as we do in Fig. 7.6, but we cannot tell anything about the IPv4 addresses themselves. This is in part due to the aggregation of the IPv4 addresses into CIDR blocks.

7.3.2 HEAT MAP

An alternate view of IPv4 addresses also uses a heatmap. In this case, we map the addresses on an xy plane. The x-axis is the 256/8s that comprise the internet. On the y-axis, we list the /16s that comprise each /8. So the x-axis is $0/8, 1/8, 2/8, \ldots, 255/8$. The y-axis is $x.0/16, x.1/16, \ldots, x.255/16$. When we pick a point n on the x-axis, the y-axis is directly related to the /16's contained in that $n/8$. The interval between n and $n + 1$ on the x-axis and $n.0/16$ and $n.1/16$ is a square. We can then use a heatmap to highlight activity in each /16. Fig. 7.7 is uses the same data as Fig. 7.6 and displays the networks routed on the Internet on November 1, 2015.

The nice thing about this heatmap as opposed to the Hilbert curve, there is not much external information required. Interpreting the Hilbert curve requires knowledge of how the curve was created and the implication of its primary property,

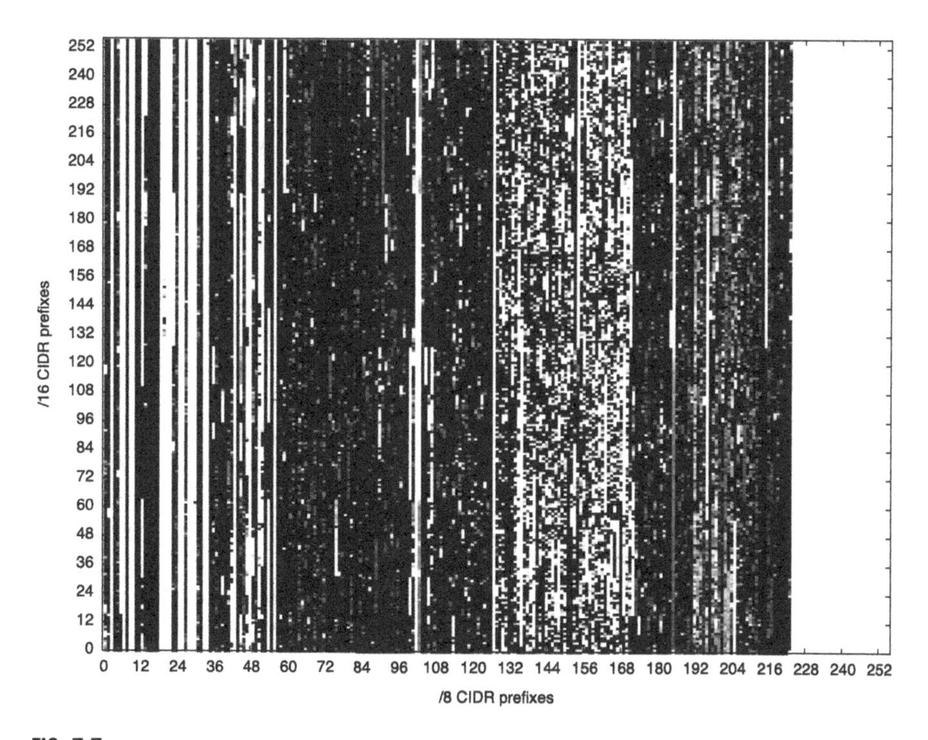

FIG. 7.7

Heatmap for IPv4 visualization.

that points close together in the xy plane tend to be closer together on the line. This map only requires the knowledge of what a /8 and a /16 are. It also lacks the useful overlay that the Hilbert curve has. As a result, similar to the Hilbert curve, this map is useful for telling a story. It is not conducive to finding anomalies or determining future research.

7.4 PLOTTING HIGHER DIMENSIONAL DATA

The human eye can see up to three dimensions. We can picture four dimensions by animating the fourth dimension so that for each value in it, we get an image. For dimensions greater than four, we have no such visualization.

Higher dimensional data is common in cybersecurity. For example, the number of bytes a sensor sees every 10 min for a day. The sensor has a vector that contains 144 elements, or a 144 dimensional vector. If we have 100 sensors, then that is a multidimensional data set. We need methods that will allow us to visualize this data for any dimension.

7.4.1 PRINCIPAL COMPONENT ANALYSIS

Principal Component Analysis (**PCA**) is a linear method to project higher dimensional data into lower dimensions. One feature of PCA is that if we decide to reverse the process, we will incur a minimum error in moving from the lower dimension to the higher dimension. It is a statistical technique that has uses beyond dimension reduction. It can be used in facial recognition and finding patterns in data as well as other applications.

Suppose we have n vectors $x_1, x_2, \ldots, x_n$ in $\mathbb{R}^p$ and we want to create n vectors in $\mathbb{R}^q$ where $q < p$ that maintain the same general structure as the data in $\mathbb{R}^p$. We compute the mean of our data set and denote it by $\bar{X}$. Now we subtract $\bar{X}$ from each vector x_i to compute x_i'. The vectors are now combined in order to create the p by n matrix X. To be precise, the elements of X are $x_{i,j}$ which are the jth element of the vector x_i'.

We will let C represent the covariance matrix of our vectors. We compute the covariance matrix of X, which has the elements $c_{i,j} = \frac{1}{n}\sum i = 1^n a_{i,j}a_{j,i}$. In matrix terms, $C = XX^T$.

We can compute the eigenvectors and eigenvalues of C. We know that each eigenvalue is associated with an eigenvector, so we can order our eigenvalues which will in turn allow us to order the eigenvectors. The largest eigenvalue is called the **principal component**. A vector in $\mathbb{R}^p$ determines a direction in $\mathbb{R}^p$ and so the principal component is the direction within the data of the greatest variation.

We want to create vectors in $\mathbb{R}^q$, so we take the first q vectors of our ordered set of eigenvectors. These can be turned into a p by n matrix by combining them in order. We will call this matrix Q. The final step is to multiply Q^T by X, which gives us a q by n matrix which we will denote by Y. Each column vector of this matrix Y is an element of $\mathbb{R}^q$. We can directly associate the ith column vector of Y with the ith column vector of X.

PCA has been used to analyze time series traffic matrices. In these matrices, the columns are the sensors and the rows are the traffic volumes measured at given time intervals. High level aggregation of traffic can mask volume anomalies, as they are buried in the normal traffic. PCA as an analysis method can find these anomalies. We are going to use PCA to visualize the traffic patterns and look for clusters.

Example 7.4.1. The Abilene connects a large number of universities and research organizations through a high performance Internet2 network. We consider traffic samples taken every 5 min over a week. From this we have gives us 2016 samples over 121 connections. We can use PCA to find like traffic patterns, as we see in Fig. 7.8.

There are two obvious clusters in the data. In other words, there are two groups of connections with similar traffic patterns.

The **Principal Coordinates Analysis** method, or **PCoA**, is related to PCA. The method uses complex linear algebra, so we will not cover the particulars here. However, it is important to note that PCoA depends on a distance matrix, as opposed to PCA. PCA requires that the multidimensional data be numerical, while PCoA does not. Any data with a metric defined can be used.

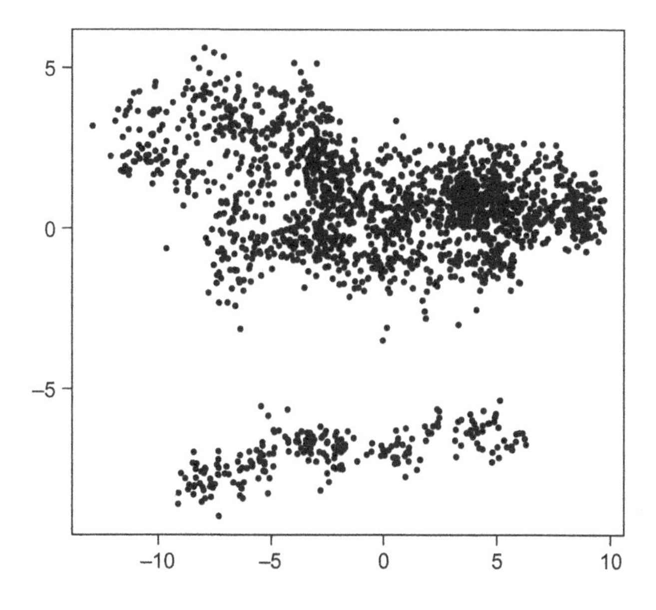

FIG. 7.8

Principal components analysis of the Abilene network.

7.4.2 SAMMON MAPPING

The Sammon Mapping, created by JW Sammon, is a nonlinear mapping for data analysis commonly used for dimension reduction. It is designed to maintain the relative distances between all of the data points in a set. This means that if there is an isolated point in the original data set, then when you apply the mapping to reduce the dimensionality of the data, the point is still isolated. Similarly, if there is a cluster of points close together in the higher dimension, that is maintained when dimensionality is reduced.

We begin with a set of higher dimensional data with n elements that we will denote as Λ. This data set must have a metric defined on it and we will denote that by $d^* : \Lambda \times \Lambda \to \mathbb{R}$. For elements $s_i, s_j \in \Lambda$ we can let $d^*_{i,j} = d^*(s_i, s_j)$ as a shorthand.

We will be using the Sammon mapping to create a new set λ that has the same number of elements as Λ. For each $s_i \in \Lambda$ there is a $s'_i \in \lambda$. We can define a bijective mapping between the two sets by mapping $f : \Lambda \to \lambda$ and defining $f(s_i) = s'_i$.

The Sammon mapping preserves relative distance. If there is an isolated element $s \in \Lambda$ then the element that corresponds to it in λ is also isolated. Similarly, if there is a cluster of elements in Λ then there will also be a cluster in λ, assuming that the cluster is found in Λ by using the metric d^*.

Eq. (7.2) is the heart of the Sammon mapping. It defines an error E which is used to determine how close the configuration of the new data points is to the configuration of the original data set. The standard procedure is to choose a set of points at random in either $\mathbb{R}^2$ or $\mathbb{R}^3$ and then compute the error. A method called gradient descent is then used to recompute the points in $\mathbb{R}^2$ or $\mathbb{R}^3$ to reduce the error:

$$E = \frac{1}{\sum_{i<j} d^*_{i,j}} \sum_{i<j}^{N} \frac{\left[d^*_{i,j} - d_{i,j}\right]^2}{d^*_{i,j}} \tag{7.2}$$

Example 7.4.2. We use the same data source as in Example 7.4.1 to create the Sammon mapping. Fig. 7.9 is the result. It is quite obvious that the method used in the Sammon mapping creates a completely different visualization and that the clusters apparent in Fig. 7.8 are not in Fig. 7.9. This is a result of the different methods of visualization. The Sammon mapping relies on a nonlinear method to find points that have similar distances from each other in the lower dimension as in the higher dimension. PCA is a linear reduction in dimension that tries to preserve as much of the structure of the higher dimensional space as it can. This also tells us that the clusters found by PCA would not be found using the Euclidean metric, as that was used in the creation of Fig. 7.9.

The original definition of the Sammon mapping uses the distances between two elements, not the elements themselves. As such, we can the Sammon mapping to visualize any data that has a metric defined on it. We can use it to visualize strings, malware, and other non-numerical data. PCA, alternatively, uses the data itself, not the distances, so it requires the data be numerical.

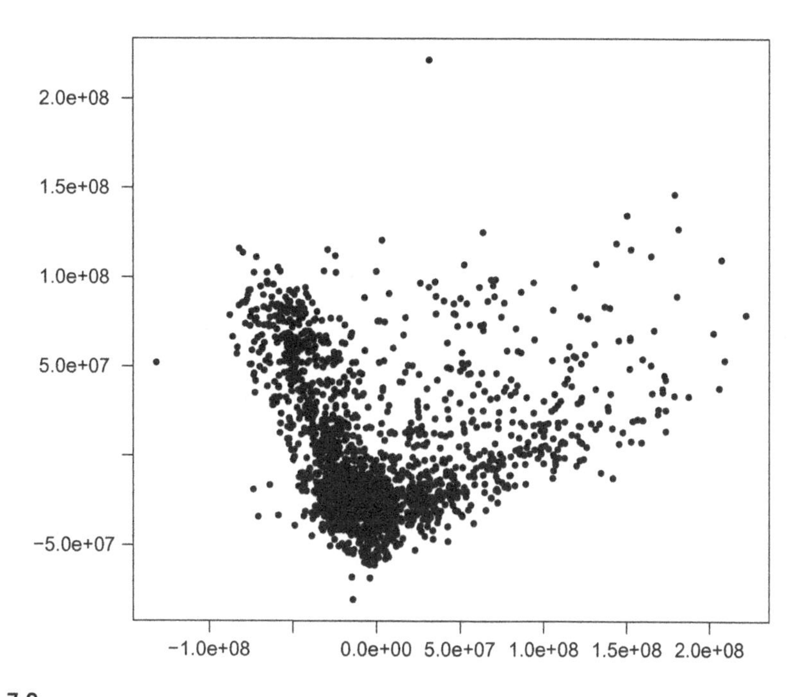

FIG. 7.9

Sammon mapping of the Abilene network.

7.5 GRAPH PLOTTING

The interesting features of graphs, such as betweenness centrality or local clustering coefficient, are not readily apparent from visualizing the graph. If a graph is a tree, that can be illustrated by a visualization of the graph. However, once the graph gets too big or complex, finding patterns is difficult. Fig. 5.1 is a visualization of a graph with 11 vertices and 11 edges. Now consider a more complex visualization of DNS in a graph. Domains have pointed to hundreds, even thousands of IP addresses, and each of those IP addresses is pointed to more domains. The visualization contains not just the vertices, but each vertex has a name. Then on top of that are the edges. So for a graph with a thousand vertices and even more edges, the visualization is very cluttered. It can be used to tell a story about the complexity of DNS, but not much else.

For a small visualization, such as Fig. 5.1, we do see that there is one vertex with a higher degree that dominates the graph. If we have a tree with a central vertex that has a degree of 100 and each of the other vertices has a degree of one, then we could see that vertex of high degree in the visualization. However, if it is not a tree, that will not be apparent in the visualization. We would have to know before examining the figure that the graph is a tree.

In short, visualizing graphs is not very useful for finding anomalies or summarizing the features of the graph. It is good for telling a story though. If we create a visualization of our network as a graph, then that visual illustrates the network at large and we can use it to explain the network. We cannot necessarily use that visualization otherwise.

There are many methods used for visualizing graphs. One is called **force-directed placement** which uses a class of algorithms designed to produced aesthetically pleasing graphs. This is not for illuminating graph features, but rather designed to create a graph that is pleasing to the eye. It is perfect for telling a story, but not designed for analysis. The goals of the method are to have edge lengths of the same size and a layout that displays symmetry.

Force directed placement uses concepts from physics to determine the layout of the graph. It considers the vertices as metal rings and the edges as equal length springs. The vertices are placed in an initial state and then let go, so that the springs move the vertices into a state that requires a minimum energy to maintain. Originally, this method was used to visualize graphs of 30–40 vertices, but the idea has been expanded so that it can be used to visualize thousands of vertices. An example of a force directed graph is in Fig. 7.10.

If we know that the graph is a tree, then there are other methods for creating the layout. In both cases, we assume that the tree has a **root**, generally a node that is designated as either the origin or destination of all other vertices in the graph.

Fig. 7.11 is an example of a classical tree layout. It emphasizes that each relationship originates at the root and is repeated to the end of the tree. There are several other ways to visualize a tree, one is the radial tree layout in Fig. 7.12. In this layout, the root is at the center of the graph and the edges radiate out in a circular fashion.

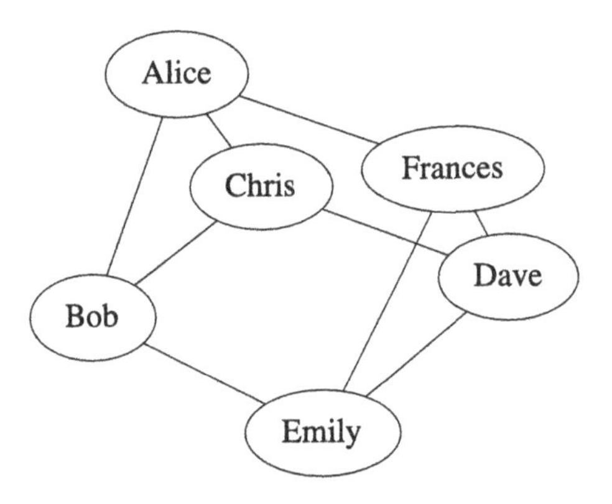

FIG. 7.10

A force-directed graph layout.

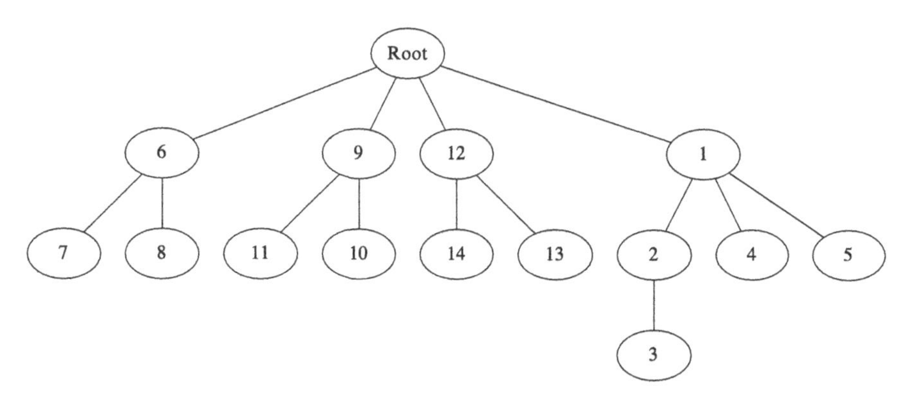

FIG. 7.11

A classical tree layout.

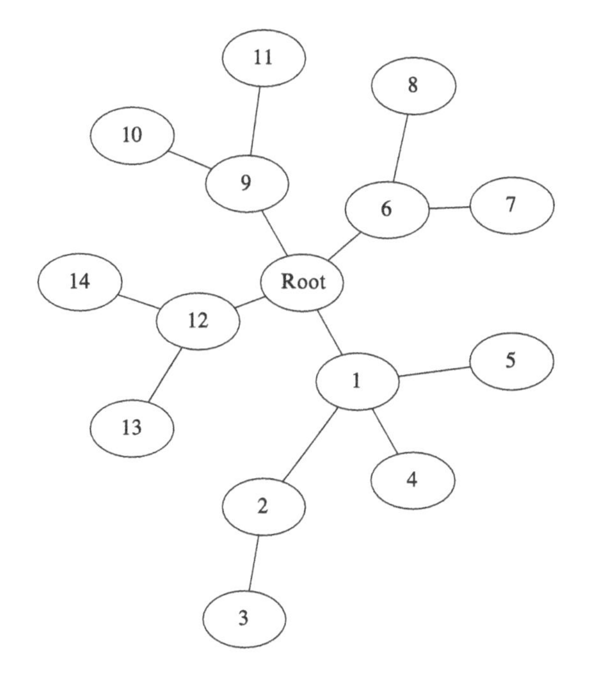

FIG. 7.12

A radial tree layout.

Both methods of visualizing trees work for graphs of reasonable size, but as the graph grows, the details are lost. If the purpose of the tree is to tell a story, such as the extensive form of a game, then the visualization is useful. The extensive form tells the story about the ways a game could be played, but that story can get lost in the complexity of the visualization.

7.6 VISUALIZING MALWARE

In Section 7.2.3 we discussed how when visualizing malware, we should also consider the story the visualization should tell. We should consider what features we want to visualize and why these features are the most useful.

One feature of malware is the functions that it calls when it is executed. If we have a set of malware, we can create a matrix of the functions called in malware versus the individual samples. Table 7.1 is such a matrix. We could use this matrix to determine common functions across malware, in hopes of tracking down the functions common to a particular malware family. This could aid us in determining the authors of the malware.

The denser the matrix, the more functions are shared among malware. We should be careful, however, with interpretation of the visualization. There are library functions which are common across most Windows applications. For example, the functions used to send and receive network data, the function used to open a socket, and the function used to determine the hosts current IP address are functions common across network applications. These are just a few of those functions, so to create an effective visualization, we need to remove the common functions from the matrix. This would allow us to highlight the functions created by the malware author. We can also use PCA on this matrix to find clusters in the data as well as to visualize it.

Another feature of malware is the section hash. These are the sub components of malware and we can hash them to find identical components across malware. We can

Table 7.1 Malware Function Matrix

AttachThreadInput	1	0	0	0	1	1	1	1	1	0
Bind	0	1	0	0	1	0	1	1	1	1
ConnectNamedPipe	1	1	0	0	0	1	0	0	1	0
CreateMutex	0	0	0	0	1	0	0	1	0	0
DeviceIoControl	1	1	1	1	1	1	1	1	0	0
EnableExecuteProtectionSupport	0	1	0	1	1	1	0	1	0	1
FindResource	0	1	1	0	1	0	0	0	1	1
FindWindow	1	1	1	1	0	1	1	0	0	0
GetAdaptersInfo	0	0	1	1	0	0	0	1	0	0
⋮										
System	1	0	1	1	0	1	0	0	0	0
VirtualAllocEx	0	1	0	1	1	1	1	1	1	1
WSASend	1	1	1	1	0	1	1	0	0	1
WSASendTo	0	0	1	1	0	1	1	1	0	0
WideCharToMultiByte	1	1	0	0	1	0	0	1	0	1
WriteProcessMemory	0	0	1	0	1	1	1	0	0	1

use a similar matrix as Table 7.1, or we can create a bipartite graph of the section hashes and the malware. The two sets that comprise the bipartite graph are the set of section hashes and the set of malware. We can then use graph plotting techniques to visualize this graph.

7.7 VISUALIZING STRINGS

Cybersecurity data includes strings. There are MD5s for malware, usernames, hostnames, we can even consider IP addresses as strings, and more. Similar to the data we have discussed so far in this chapter, we would like to be able to visualize them. This allows us to find features in the data, as we can for malware or multidimensional data.

7.7.1 WORD CLOUD

The **word cloud** is a visualization of the frequency of words in a set of texts. It first became popular in the Web 2.0 days, an early feature of websites and blogs. It was used as a navigation aid to quickly let the viewer know about the content of the website. The user could use the word cloud to quickly determine the most common topics of the blog or website. The first cloud was created in Germany in 1992, but it did not gain much traction at the time. It is still common on websites, but the popularity of the clouds has waned over time. Still, it is a quick way to visualize the common topics of a corpus.

We begin by finding the n most common words in the corpus and weighting them by frequency. The value of n is usually chosen based on the corpus. If the corpus is very short, then a high value of n would include words that do not occur often in the text. Common words such as "the," "and," and "is" are removed from the count so as not to skew the results. We want the words unique to the document, not the most common in general.

The visualization is created by using different font sizes. The largest fonts in the visualizations correlate to the most common words and the smallest correlate to the least common words. The placement of the words is done artistically, as the visualization in this case should be both informative and pleasing.

Fig. 7.13 is a word cloud created from the Wikipedia page for Internet Security, https://en.wikipedia.org/wiki/Internet_security. In this example, $n = 70$. The largest words in the visualizations are "internet" and "security." As we know the origin of the text, it is clear that these words should be the most common words in the text.

If we have this visualization for any text, then we should be able to assume the topic of the text in question, even if we do not have the text available. The visualization is created specifically to illustrate the frequency of terms in a corpus.

FIG. 7.13

A word cloud.

7.7.2 SAMMON MAPPING FOR STRINGS

In Section 7.4.2, we noted how the Sammon mapping could be used to visualize non-numerical data. As we have the Levenshtein distance on strings, we can take a set of strings and use the combination of the Levenshtein distance to create a distance matrix and the Sammon mapping to visualize the set of strings.

As opposed to the word cloud, this visualization does not extract information about a corpus. Rather, it considers the sets of words and their distance between them. We cannot use the Sammon mapping visualization to infer information about the originating text. On the other hand, we can find clusters within the original data set in the visualization.

Fig. 7.14 is a visualization of the Sammon mapping on words. We chose the words that were displayed in Fig. 7.13 to contrast the two methods of visualizations. There are no apparent clusters in this data set and at the same time, there is no way to determine information about the original text from this visualization.

On the other hand, it is a method to picture a collection of words. It allows us to create a visual model based on the strings so that we could tell a story about the strings or find anomalies. For example, if our set of strings is a collection of log messages, we could find similar messages using the Sammon mapping.

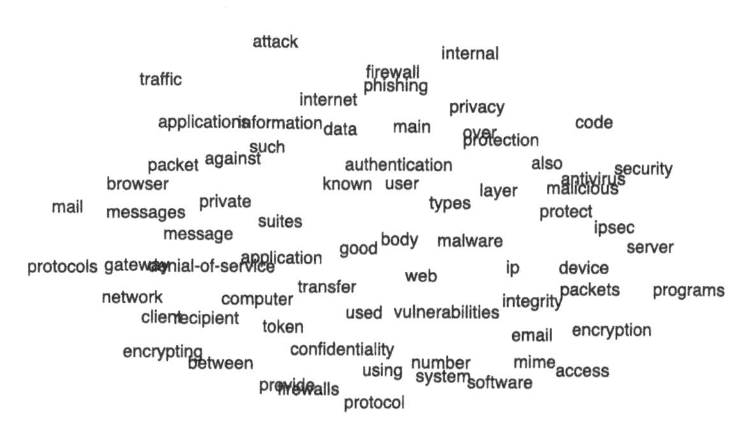

FIG. 7.14

Sammon mapping on words.

7.8 VISUALIZATION WITH A PURPOSE

As we discussed in this chapter, a visualization should have a purpose. Not only should it have a purpose, it should fulfill the purpose. In Fig. 7.1 we told a story about the ports accessed on a collection of hosts. However, it is not a clear story, it is obfuscated by the busyness of the plot. So while it could tell a story, the features of the story are not readily apparent. We can say "all ports are busy," but this isn't an interesting story. A busy installation would expect to see many ports accessed by both inbound and outbound traffic.

Instead, a single plot for each host where time is a factor is an interesting story. If there are a thousand hosts, we only examine the most critical hosts. A single visualization would tell a better story about the traffic on a host. The ports accessed over time can also be improved by including the amount of traffic, as we saw in the existence plots. Fig. 7.3 illustrated this. Again, it is for a single host, not for a collection of hosts.

The more we aggregate data, the more our story is obfuscated. It is tempting to plot all our data on one single plot, but this can muddle the story rather than clarify it. We need to be sure we choose the right data for the visualization. In Fig. 7.4 we combined time series plots to tell the story about network traffic over five hosts but we kept our story restricted to network traffic over time. By doing this we can either demonstrate the patterns of traffic over time, or we can show spikes in network traffic that could be occurring on multiple hosts. This would allow us to find network traffic anomalies that affect multiple hosts as well as allow us to summarize the network traffic across multiple hosts.

As we saw in Fig. 7.8, visualizations can be used to analyze data. We found two clusters in network traffic using PCA. This linear method demonstrated that we can visualize multidimensional data as well as analyze it. The nonlinear method for visualizing data, the Sammon mapping, also allowed us to plot the data in Fig. 7.9, but the clustering apparent in Fig. 7.8 was not apparent in the Sammon map visualization. This demonstrates that we need to consider why we are visualizing the data before we create the visualization. If we want to demonstrate the relative positions of the data to each other by dimension reduction, then the Sammon mapping is a good choice. If we want to consider the linear reduction of the data that yields a method that we can create the inverse of, then PCA is a good choice. Creating both visualizations and determining which tells our story or allows us to analyze the data is the best option.

CHAPTER

String analysis for cyber strings 8

W. Casey

8.1 STRING ANALYSIS AND CYBER DATA

A *string* is any sequence of symbols that is interpreted to represent a precise meaning. Written language, including this sentence, provides strings in which informational messages may be expressed as a sequence of words (each of which is a sequence of letters). However, natural languages such as English may give rise to ambiguous meaning. For example, consider the following statement: "Time flies like an arrow; fruit flies like a banana." In computational settings, it's important that strings (along with their encoding and interpretation) have discrete, precise meanings. Formal languages provide the general backdrop for our discussion of strings.

A central goal when analyzing cyber data is to seek a string representation for the problem's objects so that similarities in their string representations will provide a meaningful result for the analysis problem at hand. To emphasize this point, we consider signature based detection of cyber attacks and how the problem of determining safety (or that a system may be compromised) may be considered by the analysis of strings. We set the stage by providing a general background of cyber data and analysis techniques, followed by our historical examples. Then we focus the remainder of the chapter on common contemporary techniques used to analyze cyber sequential or string data.

8.1.1 CYBER DATA

Many different types of data arise from cyber security scenarios—here we will identify a few prominent data types and outline an organizational framework for thinking about cyber data. Generally, within cyber security scenarios the objects studied may or may not have much known about them. One way to think about information (known and unknown) for digital objects will be similar to that of physical objects, such as an antique. An antique is affected by a *provenance*, or a history of events, which affect its state. In the real world even a valuable historical object may have partial or disputed information concerning its provenance. For digital objects we consider provenance similarly; there can also be incomplete

Cybersecurity and Applied Mathematics. http://dx.doi.org/10.1016/B978-0-12-804452-0.00008-7

or partial awareness concerning the origin or histories of data objects (ie, files, programs, configuration settings, etc). With the notion of provenance in mind we may consider major types of cyber data.

The three main forms of data we consider are static, dynamic, and behavioral.

A *static* data analysis will focus on objects such as files, system configuration parameters, and programs (specified in a programming language or machine executable). This type of analysis may seek to identify evidence that the security of a system was breached or that a particular program is capable of breaching the system either as a direct effort (ie, malware) or indirectly (ie, a vulnerability) if attacked in a certain way. A computer system is comprised of thousands of programs and libraries, and for each of them a cyber security operator or analyst may only have a small amount of information concerning its provenance, so provenance information is usually thin thereby yielding advantages to attackers who may camouflage malware within the context of limited awareness (ie, the many system files and what they do).

A *dynamic* analysis will focus on data generated within a computer system when certain stimuli or inputs are provided. One form of dynamic data is a system log file, which contains metadata concerning the operation of various components. For example, we may monitor how a Microsoft system registry database changes before and after a given program is executed, and likewise what (if any) files are created as a consequence of executing a given program by evaluating logs or designing our own system monitors. Another and more interactive example arises when monitoring network traffic and identifying problematic communications (possibly to known command and control botnet servers). Still another example is fuzz testing in which a large variety of stimuli is provided to a program or library to find fault conditions, which may prove to be a software vulnerability. A dynamic analysis may be realized either as a system monitor or as an experiment in which the cyber analyst has created meaningful ways to observe system states.

A *behavioral* analysis will also be focused on dynamic data such as log and monitor data, but the focus of behavioral analysis will be to consider the sequences of actions and events as expressions of behaviors. Therefore, this also includes some model of behavior (such as baseline and anomalous behaviors). One type of behavioral analysis will include program tracing and a statistical model for learning a common behavior of a malware group, which distinguishes it from benign software or other malware groups. In this way, the behavioral data may include both the trace outputs as well as the model which describes a particular behavior.

8.1.2 MODES OF ANALYZING CYBER DATA

One common operational mode of analysis is a *forensic analysis* which usually takes place after an event, for example a data breach, to investigate what happened. The mode of forensic analysis is similar in nature to that of a crime scene, where an investigator and focuses on the artifacts left behind in order to gain some understanding of key questions, for example attribution (ie, who initiated the attack).

Another mode of analysis is *experimental*, often testing an object to identify how it compares to a reference data set. These types of analysis are commonly done for artifacts of unknown provenance to test if they are malware or contain vulnerabilities. Still another more operational mode is *online analysis*, and this may be thought of as a filter pipeline where actions are tested against a set of signatures in real time, with the possible outcome that a signature match may invoke a response to keep the system safe. Examples of this include network filters.

Another type of analysis mode is *formal methods and verification* and this approach is more logical in that it considers how programs or data is to be interpreted by the system and will attempt to compute or verify that certain unsafe states are not reachable. This type of analysis often employs computational processes to prove various properties of the software artifact.

Generally, in practice, various types of cyber data and analytical modes are mixed in ways to provide the best approaches for the problem at hand. In order to introduce the area of string analysis in cyber data we have selected a few of the most primitive string comparison methods which often are applied to the various data forms and as a part of a variety of modes for analyzing cyber data.

8.1.3 ALPHABETS AND FINITE STRINGS

Let Σ be a finite and nonempty set of *symbols*, also called an *alphabet*. For example, $\Sigma = \{0, 1\}$ is commonly called the *binary* alphabet, and its symbols are denoted by zero and one.

Given two symbols from Σ, that is $x, y \in \Sigma$, a juxtaposition is an operation which creates a composite object: either $x \circ y$ to emphasize that x is joined to y with an ordering of x before y, or $y \circ x$ indicating that y is joined to x with an ordering of y before x. Because $x \circ y$ may not be a symbol itself, the juxtaposition operation naturally extends the alphabet to a larger set of objects which can be created by juxtaposition between these objects or alphabet members. The limiting set of all objects created from an arbitrary number of juxtaposition operations, along with either alphabet members or other objects so created, will be the set of all strings over Σ. Objects that are created with juxtaposition can themselves be juxtaposed with other such objects, and the resulting the closure represents all such possible outcomes of objects or strings.

Following in this way, we provide a theoretical method to construct recursively the set of all strings over Σ. We will do this by creating an set of objects called Σ^k, which is the set of all strings constructible from Σ with $k - 1$ juxtapositions. We let $\Sigma^1 = \Sigma$, and next, we describe the recursion:

$$\Sigma^k = \{x \circ y | x \in \Sigma^{k-1} \text{ and } y \in \Sigma\}.$$

Notice that Σ^2 is the set of all strings created by one juxtaposition, and it includes all such single juxtapositions over all possible pairs of symbols. From the recursive form above, we see that Σ^k is constructed from $k - 1$ juxtapositions. An inductive proof would check the base case and notice that the recursive equation juxtaposes $x \in \Sigma^{k-1}$ to a symbol. Therefore there is one greater juxtaposition than the number used to create objects in Σ^{k-1}, which we can take as $k - 2$ (ie, the inductive hypothesis), thus adding together to show our claim that objects in Σ^k are created with $k - 1$ juxtapositions of symbols from Σ.

Given a string x (ie, an object created from juxtaposition), the *length* of string x, denoted $|x|$, may be defined as the first $k : x \in \Sigma^k$. Said differently, the string's *length* is the number of symbols comprising the string.

The *Kleene closure*, denoted Σ^*, is defined as

$$\Sigma^* = \cup_{k=0}^{\infty} \Sigma^k,$$

and it includes all finite strings over Σ.

A string over Σ may also be thought of as an ordered list of symbols. Here, the order of symbols is represented by natural number indices (denoted with a subscript). A string s of length n (ie, $|s| = n$) can list out the ordered sequence of symbols as $s = s_1 s_2 \ldots s_k \ldots s_n$, with the kth symbol denoted as $s_k \in \Sigma$ for $k \in \{1, 2, \ldots, n\}$.

Letting λ, ω be strings over Σ, we may make explicit the outcome of juxtaposition also called *concatenation*:

$$\lambda \circ \omega = \underbrace{\lambda_1 \lambda_2 \ldots \lambda_{|\lambda|}}_{\lambda} \underbrace{\omega_1 \omega_2 \ldots \omega_{|\omega|}}_{\omega} .$$

The kth symbol of $\lambda \circ \omega$ can be computed from the indices of λ and ω by considering the shifting of indices needed to join the head ω to the tail of λ. Letting $\phi = \lambda \circ \omega$, the kth symbol is:

$$\phi_k = \begin{cases} \lambda_k \text{ if } k \leq |\lambda| \\ \omega_{k-|\lambda|} \text{ if } k > |\lambda|. \end{cases}$$

For most string implementations and pseudo code, we use indices of nonnegative integers starting from 0 used for the first symbol of each string. Therefore, for a string of length k, the last symbol is referenced by the index $k - 1$. For a string s, we refer to the jth element as $s[j]$. A substring is a contiguous region of the sting denoted as $s[j : k]$ to indicate the string of symbols $s[j] \circ s[j + 1] \circ \cdots \circ s[k]$, meaning that it is inclusive of indices on both ends. The notation $s[j : k)$ indicates the substring $s[j : (k - 1)]$ which is not inclusive of k. A *prefix* of s is any substring of the form $s[0 : j]$, for $j \in \{1, 2, \ldots |s|\}$. A *suffix* is any substring of the form $s[k : |s|)$.

8.1.4 FORMAL LANGUAGES

The concatenation operation provides a way to construct larger strings from smaller strings. When we consider how to complete the concatenation operation on strings, we also consider the empty string, denoted as ϵ, which serves as both a left and right *idempotent* for concatenation. An idempotent leaves the object unchanged; therefore the following equations hold for all strings λ, ω:

$$\lambda \circ \epsilon = \lambda.$$
$$\epsilon \circ \omega = \omega.$$

Languages and regular expressions

A *language* over Σ is any subset of Σ^*. A *language recognizer* is a function, which given a string, decides (yes/no) if the string belongs to a certain language. Theoretical computer science is interested in the types of computing machines necessary to identify languages. One of the simplest machine models is the FSA, which may be thought of as a finite set of graph vertices called *states*; one of these vertices will have a special designation as a *start state*, and any subset of the vertices will also be designated as *accept states*. In addition, there is a *state transition rule* which is a set of labeled directional edges emanating from each state. To simplify the discussion, we continue to describe a deterministic finite automata (DFA), which is a type of FSA simpler than the non-deterministic variety. For the DFA we use an alphabet Σ to label edges. In a DFA each vertex is the source of exactly one directed edge labeled by each symbol of an alphabet. Given a string input, a DFA may decide (yes/no) to *accept* it as follows: We place a token on a vertex designated as the start state. Given a string s, we inspect one symbol at a time and apply the following rule to move the token to the next vertex: starting from the vertex with the token, we use the edge labeled with the current symbol to move the token to the next vertex (the edge is directed). Finally, after the last symbol of the string, we accept the string as a member of the language if the last vertex transitioned to is designated as an accept state.

Although simple and limited in memory (by the set of states), a DFA machine is able to recognize a language as the subset of all strings that it accepts. The types of languages that DFAs recognize turn out to be the same class as the non deterministic variety of FSA recognize, and are called the *regular languages*. Regular languages are well known to computer users who have used simple wild card substitutions (eg, `'ls *.py'` to list all the Python programs in a directory) or formed complicated *regular-expression*–based manipulation (eg, `sed 's/\(0[Xx][0-9A-Fa-f]\)/HEX-ADDRESS-FOUND:\1/g`). Known more commonly by their regular expression representations, the regular languages are a powerful and useful tool for computing. The usefulness of regular expressions is due to how they can be efficiently compiled into a DFA matcher. It is also possible to show an equivalence between regular languages, DFA recognizers, and regular expressions, therefore indicating that these are three ways of discussing the same thing. Regular expression matching is computationally very efficient and contributes to the underlying model for most malware signature-detection methods.

8.2 DISCRETE STRING MATCHING

8.2.1 HASHING

A *hash function* is a deterministic function which projects a domain of strings (or keys) into a range of bounded integers. The domain of a hash function is the space of all strings. However, with the use of an object's serialization function, we may extend the domain of a hash function to include objects, data structures, or, generally, anything which admits to an unambiguous string representation. Originally developed for information retrieval algorithms such as a hash table implementations,[1] the properties which make hash functions effective have become well known and useful for a number of computational tasks.

A particularly important consideration for a hash function concerns the distribution of range values for a randomly sampled set of keys. In particular, a hash function is said to be a *universal hash function* when the hash values of random samples tend to distribute uniformly over the range space. Notice also that the domain of a hash function is the set of all strings (infinite), and the range is a bounded set of integers (finite); therefore, the hash function will also be many to one. A *hash collision* occurs when two distinct key values map to the same integer value, and when a collision occurs in a hash table, extra steps will be required. Hash functions that are *universal* are very useful in information retrieval tasks because they can be analyzed probabilistically to understand the likelihood of hash collisions.[2] Despite the possibility of collisions, hash functions have enormous practical use because they can be computed efficiently, and in order to deal with collisions, the designer can select parameters of the hash function, the table, and the collision resolution strategy in order to tailor the efficient hashing techniques to particular problems.

Closely related to uniform hash functions are *cryptographic hash functions*, which are functions designed to be difficult to invert. A type of cryptoanalysis technique called *frequency analysis*, which studies a cryptographic function statistically, can be used to find the weaknesses of a cryptographic hash function. Therefore, in order to be secure, a cryptographic hash function must also be a *universal hash function*. Cryptographic hashes such as MD5, SHA256, and SHA1 are commonly used to perform integrity checks of communicated messages. The wide-scale use of cryptographic hash functions has also taken root in the security community where hash functions are also used to identify artifacts. It is common to refer to malware, binary artifacts, and library and reference objects by the 32 hex digits comprising its MD5 sum.

[1] A hash table can be used to implement an associative array for organizing a key-value store where a user may wish to add and remove data dynamically.

[2] A hash collision occurs when two keys map to the same output value.

Birthday party and universal hash function collisions

The range of a hash function is a finite (bounded) set of integers R, but the domain is an infinite set of strings. Therefore hash collisions are possible, and among a set of n objects, there is some probability that any two of them will have a common hash value. For example, if n is greater than $|R|$, a hash collision is guaranteed (eg, with probability 1) by the pigeon hole principle. Given a universal hash function, the question of a collision's likelihood among n objects arises naturally. This problem is widely known in probability as *the birthday problem*. Among n peoples we ask what is the probability that any two of them have the same birthday? A related question is how many people n do we need before the probability that two birthdays collide exceeds a certain value?

To consider the possibilities of a hash collision for universal hash functions we ask the same question for a set of n keys, or strings, or binary artifacts. Lets fix MD5 as the universal hash function under consideration. Under the assumption that the function is universal (ie, the outputs are uniformly distributed), the probability that any particular string hashes to a given value may be considered to be $\frac{1}{2^{64}}$. The probability that any pair of n strings hash to the same value can be considered with its complement event; that is, given n distinct strings, we wish to know the probability that the hash values are all distinct (ie, no collisions). We compute this probability in the context of universal hash functions for binary artifacts.

Let us assume that we have a random sample of n distinct binary strings (artifacts) and a universal hash function. Given an arbitrary order of binaries, let event E_k describe the event that the kth binary artifact hashes to a value different from each of the previous distinct $k-1$ hash values:

$$\text{Prob}(E_k) = \text{Prob}(E_{k-1}) \times \underbrace{\left(1 - \frac{k-1}{2^{64}}\right)}_{\text{Avoids previous values}}.$$

With the obvious base case that Prob $E_1 = 1$, letting R be the range of the universal hash function, we obtain the probability that, among n artifacts, all n binaries will hash to differing range values with probability:

$$\text{Prob}(E_n) = \prod_{k=1}^{n} \left(1 - \frac{k-1}{|R|}\right). \tag{8.1}$$

Additionally, we may be interested in the birthday problem; that is, for what value of n will $\text{Prob}(E_n) < \kappa$, implying that, with probability greater than κ, a collision is found. We now present an approximation method used to solve for such an n. To approximate these probabilities, we consider the product of exponents

$$P_n = \prod_{k=1}^{n} \exp\left(-\frac{k-1}{|R|}\right),$$

and we notice that the inequality $1 - x \leq \exp(-x)$, for all $x \in [0, 1]$, can be applied term by term to bound Prob(E_n) by P_n:

$$\text{Prob}\,(E_n) < P_n, \text{ for every } n > 0.$$

Therefore, with this bound, $P_n < kappa$ indicates that Prob(E_n) $< \kappa$.

Notice that P_n can also rewritten with an exponential sum:

$$P_n = \exp - \left(\frac{1}{|R|} \sum_{i=1}^{n} (k-1) \right) = \exp\left(-\frac{n(n-1)}{2|R|} \right).$$

For example, the bound implies that the probability of a hash collision exceeds $\frac{1}{2}$ when n is sufficiently large that $P_n < \frac{1}{2}$ or

$$\exp\left(-\frac{n(n-1)}{2|R|} \right) < \frac{1}{2},$$

$$\left(-\frac{n(n-1)}{2|R|} \right) < \log\frac{1}{2}, \text{ and}$$

$$n(n-1) > 2|R| \log 2. \tag{8.2}$$

Therefore, the minimum value of n that satisfies Eq. (8.2) also implies that a hash collision occurs with probability $\frac{1}{2}$ or greater. To approximate such a minimum value for n, we approximate the left hand side of Eq. (8.2) as n^2 and solve the following test equation:

$$n^2 = 2|R| \log(2). \tag{8.3}$$

When $|R|$ is itself a power of 2, say $|R| = 2^w$, the solution to the test equation (ie, Eq. (8.3)) is $n = 2^{\frac{w+1}{2}} \log(2)$. Therefore, the probability of a hash collision for MD5 (where $w = 64$) exceeds $\frac{1}{2}$ when $n \approx 2^{32.5} \log(2)$ or when n is around 4.2 billion objects. Starting from this value of n, we can determine more a accurate minimum value for n; however, the described bounds and approximations help us to obtain an estimate quickly.

Determining a hash collision itself may amount to an important finding for cybersecurity. For cryptographic hash functions, the ease with which a hash collision can be found or constructed may be exploited to subvert the integrity of a message. Generally, an application which uses a universal hash function will also consider the probability of collisions which is guaranteed when the input space is infinite and range values are bounded.

8.2.2 APPLICATIONS OF HASHING

With this understanding of hash functions and their inherent limitations due to hash collisions themselves due to hash functions' finite range, we next focus on how their efficiency can be used to study strings that are relevant in cybersecurity.

Hash functions are efficient when identifying matching strings. Given that data and code objects may be represented by strings, a hash function can be used to match objects as well. Take, for example, a binary executable of unknown origin as M. Often, the executable file format includes a natural division of code and data into sections, segments, or pages which we may use to represent M as a list of constituent string objects $M = [m_1, m_2, \ldots, m_K]$. Moreover these constituent strings may often be readily broken down further into data (as known data structures are simple to extract) or code (sections may be have a listing of functions making their extraction straightforward). Although further resolution of a data object into constituent objects is typically very powerful and revealing, for illustrative purposes, we need only consider the first-order division of M into constituent strings $[m_1, \ldots, m_K]$.

An adversary, aware that hashing techniques may be used to identify provenance in malware, may attempt to obscure the provenance information, but this extra effort will be costly due to the typical software pattern of code reuse. This inherent tradeoff faced by the adversary gives rise to the possibility that provenance may be recovered from binary artifacts such as malware. To detect the presence of code reuse with hash functions, we present several techniques below.

Bag of hashes, or bag of numbers

Given a universal hash function U (a mapping $\Sigma^* \rightarrow R$), and an artifact naturally divided into constituent strings $M = [m_1, m_2, \ldots, m_K]$, we may use U to derive a set called a *bag of numbers* representing the artifact as $\mathcal{R}_U(M)$:

$$\mathcal{R}_U(M) = \{U(m_1), U(m_2), \ldots, U(m_K)\}.$$

Using hash function U with range R, an object comprised of K constituent objects is therefore mapped into a set of R^K. This set, the bag of numbers, can be compared with other bags derived from reference objects. Bags can be compared with the Jaccard coefficient to measure similarity to previously known objects.

An important caveat to this analysis is the possibility of hash collisions which would introduce a false sense of similarity. For this reason it's important to understand the design goals and properties of the employed hash function U and under what conditions hash collisions become likely.

This technique may be applied in the study of portable document format (PDF) based malware. The PDF format defines a tree of constituent objects and stores these objects as streams (serialized representations). So, letting U be MD5, we can consider the bag of numbers to identify common objects across multiple PDF files. Similarly, binary executables are comprised of sections, segments, and pages; therefore, executables in a given file format may be compared using bag of numbers. Of particular interest are the strings which express the functions, procedures, and data objects within an executable. Although function specification strings may be made more difficult to extract from code, they can also be observed during runtime by monitoring the execution trace.

Normalizing bag of numbers

In many settings the use of a hash function is overly sensitive to small but unimportant modifications within a set of related strings. While we later present techniques which inherently address small string modifications, we now consider techniques which compensate for the overly sensitive universal hash functions, thus retaining the efficiency of hashing. This simple augmentation of the bag of numbers can broaden its use and also focus its precision tremendously. The technique presented here can be thought of as a simple bag of words that employs a string-rewriting function ϕ prior to the application of the hash function U. The string-rewriting function can modify strings to reduce variance either by the removal of noisy elements or the presentation of a string in a standard or canonical form in a process we refer to as *normalizing*.

Given a universal hash function U (a mapping $\Sigma^* \rightarrow R$), a string-rewriting function $\phi : \Sigma^* \rightarrow \Sigma^*$, and an artifact naturally divided into constituent strings $M = [m_1, m_2, \ldots, m_K]$, we may use U to derive a set called a *normalized bag of numbers*, representing the artifact as $\mathcal{R}_{U \circ \phi}(M)$:

$$\mathcal{R}_{U \circ \phi}(M) = \{U(\phi(m_1)), U(\phi(m_2)), \ldots, U(\phi(m_K))\}.$$

We next illustrate these techniques on two functions listed in Table 8.1 and extracted from a binary executables. The functions are specified in assembly language and are not identical, but they are structurally similar and functionally related. Letting U be the MD5 hash function, we consider two functions ϕ_{PIC} and ϕ_{RIC}, which will provide a means to compute a position-independent code (PIC) normalized bag of numbers, as well as a a register-independent code (RIC) normalized bag of words. The selected functions show a case where similarities and matching provide meaningful conclusions concerning related functions and come with few data manipulations in addition to the basic hash function.

To illustrate our string-rewriting functions ϕ_{PIC} and ϕ_{RIC}, we select the function labeled ISPUNCT, and for each line of code, we attribute positional data (second column) and register data (third column). These attributes are used to group lines of code into sets with a common attribute defining the method we use to split function into constituent parts. Further, while rewriting each line of code, we replace each item with a parameterized string (Table 8.2).

For each line of code, the function ϕ_{PIC} can be executed by identifying all hexadecimal numbers (using a regular expression) and extracting them into a buffer. The distinct extracted numbers, in left to right order, as found in the line of code, become the attribute for the line of code. Next, to rewrite the line of code, each distinct hexadecimal number is replaced with a string. For example, we replace the jth distinct hexadecimal number with string $P.j$ everywhere it occurs in the line of code. Next, the rewritten lines of code are reordered first by attributes and second by original index. Finally, the strings with a common attribute are concatenated into a constituent string. So, for example, function ISPUNCT is rewritten and grouped by attribute in Table 8.3; therefore, the function ISPUNCT will be represented by

Table 8.1 Two Functions Expressed in Assembly and Extracted From the Program "gawk"

Ispunct	**Isspace**
pushq %rbp	pushq %rbp
movq %rsp, %rbp	movq %rsp, %rbp
cmpl $127, %edi	cmpl $127, %edi
ja 0x100021f0f	ja 0x100021bcf
movslq %edi, %rax	movslq %edi, %rax
movq 229669(%rip), %rcx	movq 230501(%rip), %rcx
movl 60(%rcx,%rax,4), %eax	movl 60(%rcx,%rax,4), %eax
shrl $13, %eax	shrl $14, %eax
andl $1, %eax	andl $1, %eax
popq %rbp	popq %rbp
ret	ret
movl $8192, %esi ; 0x100021f0f	movl $16384, %esi ; 0x100021bcf
callq 0x10004bd7a	callq 0x10004bd7a
testl %eax, %eax	testl %eax, %eax
setne %al	setne %al
movzbl %al, %eax	movzbl %al, %eax
popq %rbp	popq %rbp
ret	ret
nopw %cs:(%rax,%rax)	nopw %cs:(%rax,%rax)

Functions ISPUNCT *and* ISSPACE *exhibit a high degree of similarity in their structure. The callq commands indicate that flow of control is transferred to differing locations where specific actions are performed, and the brevity and similarity of these functions indicate that they may be wrappers or templates. Notice there are only a few differences in lines of code, including differing jump locations at line 4 represented in hexadecimal notation.*

$[f_0,f_1,f_2]$ with f_0 being the concatenated code of the first 17 lines of Table 8.3 and refer to the lines of ISPUNCT without hexadecimal numbers. On the other hand, f_1 is line `ja P.0`, and string f_2 is the string `callq P.0`. The bag of numbers for function ISPUNCT, normalized with ϕ_{PIC}, is shown in Table 8.5.

For each line of code, the function Φ_{RIC} can be executed by identifying all register addresses and extracting them into a buffer. The distinct register addresses, in left to right order, as found in the line of code, become the attribute for the line of code. Next, to rewrite the line of code, each distinct register address will be replaced with a string: we replace the jth distinct register address with string $P[j]$ everywhere it occurs in the line of code. Next, the rewritten lines of code are reordered first by attributes and second by original index. Finally, the strings with a common attribute are concatenated into a constituent string. So, for example, function ISPUNCT is rewritten and grouped by attribute in Table 8.4; therefore, the function ISPUNCT is represented by $[f_0,f_1,f_2,\ldots,f_{13}]$, with f_0 being the concatenation of the first four

Table 8.2 The Function ISPUNCT With Attributes

	ISPUNCT (Lines of asm)	Positional Attribute	Register Attribute
1	pushq %rbp	0	{rbp}
2	movq %rsp, %rbp	0	{rsp, rbp}
3	cmpl $127, %edi	0	{edi}
4	ja 0x100021f0f	0x100021f0f	{}
5	movslq %edi, %rax	0	{edi, rax}
6	movq 229669(%rip), %rcx	0	{rip, rcx}
7	movl 60(%rcx,%rax,4), %eax	0	{rcx, rax, eax}
8	shrl $13, %eax	0	{eax}
9	andl $1, %eax	0	{eax}
10	popq %rbp	0	{rbp}
11	ret	0	{}
12	movl $8192, %esi ; 0x100021f0f	0	{esi}
13	callq 0x10004bd7a	0x10004bd7a	{}
14	testl %eax, %eax	0	{eax , eax}
15	setne %al	0	{al}
16	movzbl %al, %eax	0	{al, eax}
17	popq %rbp	0	{rbp}
18	ret	0	{}
19	nopw %cs:(%rax,%rax)	0	{cs, rax , rax}

Each line of code is attributed with various values depending on content. The positional attribute column identifies a hexadecimal address (if one exists within the line of code) or otherwise evaluates to zero. The register attribute column identifies the set of registers implemented for each line of code; these values are indicated with a set whose default value is an empty set.

lines in Table 8.4. The bag of numbers for function ISPUNCT, normalized with ϕ_{PIC}, is shown in Table 8.6.

Notice that the resulting set of integers, represented in Tables 8.5 and 8.6 as hexadecimal strings, are independent of the attribute values and indicate the bag of numbers for code-independent code groupings and register-independent code groupings.

With these two normalized bags of words described, we return to the problem of comparing related functions. In this case, ISPUNCT and ISSPACE shown in Table 8.1 are mapped to the PIC-normalized bag of numbers:

$$\begin{aligned}\mathcal{R}_{\text{MD5}\circ\phi_{\text{PIC}}}(\text{ISPUNCT}) &= [8853573ca3512634642a5f574d1df63a,\\ &\quad 2dc4564e7ac59eb6c0ab4fb9aff26bbc,\\ &\quad 9679a36898b4fd96f15936295ea146b5], \text{ and}\\ \mathcal{R}_{\text{MD5}\circ\phi_{\text{PIC}}}(\text{ISSPACE}) &= [a612737df118f1dd73a5dd88fd78ba7e,\\ &\quad 2dc4564e7ac59eb6c0ab4fb9aff26bbc,\\ &\quad 9679a36898b4fd96f15936295ea146b5].\end{aligned}$$

Table 8.3 Address-Sensitive Regrouping of Function Contents for Hashing

ISPUNCT (Lines Organized by Position)	Positional Attribute
pushq %rbp	
movq %rsp, %rbp	
cmpl $127, %edi	
movslq %edi, %rax	
movq 229669(%rip), %rcx	
movl 60(%rcx,%rax,4), %eax	
shrl $13, %eax	
andl $1, %eax	
popq %rbp	0
ret	
movl $8192, %esi ; 0x100021f0f	
testl %eax, %eax	
setne %al	
movzbl %al, %eax	
popq %rbp	
ret	
nopw %cs:(%rax,%rax)	
ja P.0	0x100021f0f
callq P.0	0x10004bd7a

Notice the parameterized replacement of attribute values, thereby normalizing away the variability but maintaining the a structural meaning.

The PIC-normalized bags of numbers between the two functions have two numbers (out of three) in common.

We now display a shortened form of the RIC-normalized bag of numbers:

$$\mathcal{R}_{\mathrm{MD5}\circ\phi_{\mathrm{RIC}}}(\textsc{IsPunct}) = [1636.48f1, 1179.62ff, 01f7.eb04, f377.9eaf, ef3a.26ab, 0fba.128f, ca2a.08c0, 1760.b11e, 9464.ef9f, 2991.8ea5, 8eaf.2a6f, 525d.874b, 54cf.176a], \text{ and}$$

$$\mathcal{R}_{\mathrm{MD5}\circ\phi_{\mathrm{RIC}}}(\textsc{IsSpace}) = [b9d5.c06f, 1179.62ff, 01f7.eb04, f377.9eaf, bf08.8804, 0fba.128f, ca2a.08c0, 1760.b11e, b7e0.efb6, 2991.8ea5, 8eaf.2a6f, 9dd5.59a2, 54cf.176a].$$

The RIC-normalized bag of numbers for the two functions has nine numbers (of 13) matching. Both results show strong similarities, and more importantly, we illustrate with this example how an analyst (familiar with various data/executable file formats) may use hash functions to great effect by combining them with data

Table 8.4 Regrouping Lines of Code by the Register Groupings

ISPUNCT (Lines Organized by Position)	Register Attribute
ja 0x100021f0f	
ret	{}
callq 0x10004bd7a	
ret	
setne P[0]	{al}
movzbl P[0], P[1]	{al, eax}
nopw P[0]:(P[1],P[1])	{cs, rax , rax}
shrl $13, P[0]	{eax}
andl $1, P[0]	
testl P[0], P[0]	{eax , eax}
cmpl $127, P[0]	{edi}
movslq P[0], P[1]	{edi, rax}
movl $8192, P[0]	{esi}
pushq P[0]	
popq P[0]	{rbp}
popq P[0]	
movl 60(P[0],P[1],4), P[2]	{rcx, rax, eax}
movq 229669(P[0]), P[1]	{rip, rcx}
movq P[0], P[1]	{rsp, rbp}

The function contents are reorganized for hashing. Notice the parameterized replacement of registers, thereby normalizing away the variable while maintaining the meaning with the attribute known.

Table 8.5 ISPUNCT Hashes by Positional Value

Positional Attribute	Hash Value
0	8853573ca3512634642a5f574d1df63a
0x100021f0f	2dc4564e7ac59eb6c0ab4fb9aff26bbc
0x10004bd7a	9679a36898b4fd96f15936295ea146b5

normalization functions ϕ that remove the unimportant variations in data to induce a common hash. In this way, the technique may be extended generally to many problems.

Cryptographic hashing and universal hash functions are simplistic, efficient, and useful for digesting large and complex data objects into a bag of numbers where they can be compared to a reference set. For large data sets, it's important to understand the properties of the underlying universal hashing properties. With additional knowledge of how unimportant variations in data may be normalized, the normalized bag of words offers endless possibilities to capture similarity signals in data. However, it's not always the case that we have knowledge of how our data should be interpreted,

Table 8.6 ISPUNCT Hashes by Register Use Attribute

Register Attribute	Hash Value
{}	1636e4a1b173928d704847bde7ec48f1
{al}	11790bdd714acb69d25005272c6562ff
{al, eax}	01f7993fcda73368dbd21eb36ce8eb04
{cs, rax, rax}	f377c762ce0a6ca886b668696f659eaf
{eax}	ef3a597d462aaa438580218f89bd26ab
{eax, eax}	0fba84064444a0580c0c30844bc9128f
{edi}	ca2a4f537d823ddedaa58baa6ad908c0
{edi, rax}	1760137b565cff8d0362f3ffec48b11e
{esi}	9464fee12647af58e837fb0b7a04ef9f
{rbp}	29912154aa22885ccb685f10ff5b8ea5
{rcx, rax, eax}	8eaf278cbeac5b50765b97d9e5842a6f
{rip, rcx}	525db9fa4ef126036667343cbd94874b
{rsp', rbp}	54cf2569d9d48a75a2bb76131e29176a

and for this reason, it's important to consider more generally methods for reasoning about string similarity with few assumptions. We conclude this section with a discussion of several other techniques that have various degrees of implementation in cybersecurity practice.

8.2.3 OTHER METHODS

Fuzzy hashing techniques include a large number of approximate techniques, including *context-sensitive hashing*, which may be considered a means to derive natural boundaries in a string by use of context clues within the string. Another common hashing technique is Bloom filters, often used to determine set membership.

A Bloom filter is a randomized algorithm for testing whether a given object is a set member, and it is built on top of a set of k independent hash functions (recall each hash function outputs a bounded integer) which establish a set of bits to set within a fixed-width byte array. Therefore, if we are checking set membership, we can determine the bits which would necessarily be set (assuming the object is a member of the set), and with this, we have a way of accurately refuting the set membership and checking set membership probabilistically.

Another method which introduces a modulus of continuity to a hashing function is to consider the histogram or counts over various symbols in the alphabet. These values have many of the desired reduction properties of a hash function, and they are sensitive to slight variations arising from string mutations.

Suppose that Σ is the alphabet with k symbols. Given a string s of length n, let

$$c_j = \sum_{i=1}^{n} \delta(s[i], \Sigma_j),$$

where $\delta(x, \Sigma_j)$ is 1 if a given symbol x is equal to the jth symbol from Σ and is 0 otherwise. The histogram hash is $H : \Sigma^* \rightarrow [0,1]^k : s \rightarrow \langle \frac{c_j}{n} \rangle_{j=1}^k$, where $n = |s|$. This hash can be used to create a k-dimensional probability vector projection of a given string. When a string s is modified to string s' via a few mutations or otherwise minor changes, we can expect $H(s), H(s')$ vectors to be close in distance as well. A final heuristic method which may prove to be very useful for the analysis of cyber data is the method called winnowing which was developed for detecting code plagiarism.

Additionally, many subgraph matching and approximate matching techniques exist for analyzing control flow structures in code, and we forego the discussion of those methods here.

8.3 AFFINE ALIGNMENT STRING SIMILARITY

In some cybersecurity-data–driven problems, we wish to reason about data for which we may have no or little information concerning how it is to be formally interpreted. In this scenario, we may be analyzing a firmware program within a poorly documented file format, or we may be reverse-engineering a program that uses some nonstandard data structures and function calling conventions. Often times these problems may focus on finding patterns which indicate data structures, objects, or the entry/return structures of function calls. To consider these problems, we generally imagine strings with common prefixes and suffixes but having interesting forms of variations in-between, and to address this we broaden our approach to consider string-alignment techniques from biology. In this section, we consider several important measures from bioinformatics and their application to code comparison by focusing on the Needleman-Wunsch global affine alignment algorithm. We summarize the common optimization technique known as dynamic programming which underlies the major string similarity measures, including Levenshtein distance, Needleman-Wunsch, and Smith-Waterman similarities. The underling optimization concept is summarized in the *principle of optimality* which is used to construct dynamic programming implementations for optimal alignments. We construct the Needleman-Wunsch alignment algorithm to illustrate dynamic programming and show how it may be used to explore the functions of a binary program, such as would be necessary in a reverse-engineering task.

8.3.1 OPTIMALITY AND DYNAMIC PROGRAMMING

In optimization problems where a solution can be built or synthesized from the solutions of smaller subproblems, a particularly interesting property has been discovered called *the principle of optimality*. As an example of a problem whose solution can be synthesized from like or similar subproblems, consider the problem of determining a minimal distance path through a graph starting from a vertex A and ending at vertex Z. Because a minimal distance path may itself go through some intermediate point (P), we may ask how the problem of minimizing a path between any of the two endpoints and the intermediate point P may relate to the problem of minimizing a path from A to Z. Naturally, these problems are directly related in that the optimal solution, the minimal distance path from A to Z, will also have to be constructed from optimal paths of subproblems in particular for any pair of verities named in the solution. Let $Opath(X, Y)$ be the optimal path from vertex X to vertex Y, listed as a string over vertices. If the optimal solution from A to Z computed as $Opath(A, Z)$ includes a vertex P, then we reason that there is a prefix relation among the solutions to subproblem $Opath(A, P)$ and $Opath(A, Z)$. This reasoning maybe supported by exploring the logical possibility that P is found in $Opath(A, Z)$ and the optimal path $Opath(A, P)$, somehow doesn't match a prefix of $Opath(A, Z)$, then would there not be a more optimal path from A to Z? The previous argument can form a proof by contradiction that the optimal solutions must be a synthesis of solutions to related subproblems.

When an optimization problem can be seen as a synthesis of optimal solutions for subproblems, the principle of optimality applies. The principle of optimality leads to a solution in the form of dynamic programming where solutions are constructed for the most trivial subproblems first and those solutions are extended with branch and bound to larger problems.

8.3.2 GLOBAL AFFINE ALIGNMENT

We begin by describing the problem and related subproblems. We then show the principle of optimality and define a recursion to solve the problem as a Needleman-Wunsch algorithm. Given two lists of integers L_1, L_2 of length n, m, we imagine an edit procedure which turns L_1 into L_2. We let the alphabet Σ be a bounded set of integers including all the symbols of L_1 and L_2. We envision a cursor position in both L_1 and L_2 at position (i, j), with $0 \leq i < n, 0 \leq j < m$, and for each cursor position, one of three edits operations can be performed:

- *Mutate*, modifying $L_1[i]$ to match that of $L_2[j]$ and increment i, j
- *Insert*, moving cursor i ahead one position leaving j constant
- *Delete*, moving cursor j ahead one position leaving i constant

When a mutate operation is performed, a cost depending on the two characters will be assessed. All costs are parameterized by a substitution matrix $S : \Sigma \times \Sigma \rightarrow \mathbb{R}$ as $S(L_1[i], L_2[j])$. After a mutate operation, the cursor will be updated from (i, j) to $(i + 1, j + 1)$.

When an insert operation is performed, it will cost either ϵ, if an insert was previously performed, or otherwise δ. This differing cost is intended to model a onetime cost of δ (usually large in magnitude) for initializing a cut-and-paste and cost ϵ (usually smaller in magnitude) for extending the cut-and-paste patch of symbols. Together, these two parameters provide a linear or affine function $y = \epsilon x + \delta$ which gives our problem its name. Notice that, in Levenshtein distance, we would have $\epsilon = \delta$.

Similarly, when a delete operation is performed, it will cost either ϵ, if a delete operation was previously used, or δ otherwise. A deletion maybe thought of as an insertion for the other string.

Given the two strings and these edit operations, the problem of *global affine alignment* is to modify $L_1[i]$ into $L_2[j]$ with a maximum reward. To be clear, we will have $\delta < \epsilon < 0$ as cost penalties, and $S(\sigma,\sigma) \geq 0$ for $\sigma \in \Sigma$. Further, we will require that S is symmetric and $S(\sigma,\sigma) \geq S(\sigma,\lambda)$ for any $\sigma, \lambda \in \Sigma$. The solution will be comprised of both a score which is optimal and a sequence of edit operations which will modify L_1 into L_2 in order to attain the optimal score.

To organize subproblems, we consider the substrings of L_1, L_2 as $L_1[a_1 : b_1]$ and $L_2[a_2 : b_2]$ with $a_i \leq b_i$. To establish the principle of optimality, we need to show that, if an optimal global affine alignment between L_1, L_2 also forms an alignment between substrings $L[a_1 : b_1]$ and $L[a_2 : b_2]$, then the alignment of $L[a_1 : b_1]$ and $L[a_2 : b_2]$ must also be optimal. This optimization principle can be established with a proof by contradiction argument similar to that of the minimal path problem.

In Algorithm 8.1 and supporting routines Algorithms 8.2–8.4, we outline a global affine alignment algorithm known as Needleman-Wunsch for any two sequences of integer values L_1 and L_2. Throughout, the algorithms will assume a setting for δ and ϵ and S.

ALGORITHM 1 NEEDLEMAN-WUNSCH ALGORITHM

Data: Given L_1, L_2 ordered lists of integers of length n, m, parameters ϵ, δ, and S are given.
Result: Global alignment similarity score,
`InitializeBoundaryConditions`(m,n) % to initialize M, E, F, P_M, P_E, P_F
for (i,j) *in* `IndexOrder` (m,n) **do**

$v_M = \max(M[i-1,j-1], E[i-1,j-1], F[i-1,j-1])$

$$P_M[i,j] = \begin{cases} (i-1,j-1) \text{ if } v_M = M[i-1,j-1] \\ (i,j-1) \text{ if } v_M = E[i-1,j-1] \\ (i-1,j) \text{ if } v_M = F[i-1,j-1] \end{cases}$$

$M[i,j] = v_M + S(L_1[i-1], L_2[j-1])$ % Similarity contribution

$v_E = \max(M[i-1,j] - \delta, E[i-1,j] - \epsilon, F[i-1,j] - \delta)$

$$P_E[i,j] = \begin{cases} (i-1,j-1) \text{ if } v_E = M[i-1,j] - \delta \\ (i,j-1) \text{ if } v_E = E[i-1,j] - \epsilon \\ (i-1,j) \text{ if } v_E = F[i-1,j] - \delta \end{cases}$$

$E[i,j] = v_E$

$v_F = \max(M[i,j-1] - \delta, E[i,j-1] - \delta, F[i,j-1] - \epsilon)$

$$P_F[i,j] = \begin{cases} (i-1,j-1) \text{ if } v_F = M[i,j-1] - \delta \\ (i,j-1) \text{ if } v_F = E[i,j-1] - \delta \\ (i-1,j) \text{ if } v_F = F[i,j-1] - \epsilon \end{cases}$$

$F[i,j] = v_F$

end
return $\max(M[n,m], E[n,m], F[n,m])$

At the core of Needleman-Wunsch algorithm are the decisions about which edit history advances the cursor position from $(0,0)$ to (i,j) in the optimal manner. At each cursor position, various optimal histories are considered in correspondence with the three edit operations. Notice that Algorithm 8.3 specifies that the cursor positions are explored as a wavefront starting from $(0, 0)$, and in such a way that prior to the exploration of (i, j), M, E, and F are computed for cells $(i-1, j-1), (i, j-1)$, and $(i-1, j)$. At each cursor position, Algorithm 8.1 explores edit histories, but advances the histories that are scoring optimally, thereby implementing a branch-and-bound strategy cutting out the considerations of a vast number of edit histories at each step. The algorithm cuts these suboptimal possibilities because of the principal of optimality which essentially guarantees that the histories which are cut cannot in some way contribute to an overall optimal solution to the full problem.

To initialize the recurrence, the following boundary values are set in Algorithm 8.2.

ALGORITHM 2 INITIALIZE BOUNDARY CONDITIONS

Result: Initialize data M, E, F, P_M, P_E, P_F for Needleman-Wunsch.
Data: Given lengths m, n, and parameters δ, ϵ,
$M = Zeros(n, m), E = Zeros(n, m), F = Zeros(n, m)$,
$P_M = Zeros(n, m), P_E = Zeros(n, m), P_F = Zeros(n, m)$
for i *in* $[1, \ldots, (n+1)]$ **do**
  $E[i, 0] = -\delta - \epsilon(i-1)$
  $P_E[i, 0] = (i-1, 0)$% Set the back-pointer
  $F[i, 0] = -\infty$
  $M[i, 0] = -\infty$
end
for j *in* $[1, \ldots, (m+1)]$ **do**
  $F[0, j] = -\delta - \epsilon(j-1)$
  $P_F[0, j] = (0, j-1)$ % Set the back-pointer
  $E[0, j] = -\infty$
  $M[0, j] = -\infty$
end
$M[0, 0] = 0$
$B[0, 0] = 0$
$E[0, 0] = -\infty$
$F[0, 0] = -\infty$

To create an ordering over the array that guarantees that M, E, and F are computed for cells $(i-1, j-1), (i, j-1)$, and $(i-1, j)$ prior to (i, j), we create a wavefront ordering using the following Algorithm 8.3. This ordering can be used in all of the dynamic programming alignment algorithms.

ALGORITHM 3 INDEX ORDER

Data: Given lengths m, n.
Result: An array index order or itinerary listing indices for dynamic programming.
$order = []$
for d *in* $[1, \ldots, (m+n+1)]$ **do**
  for s *in* $[\max(1, d-(m-1)), min(d, n)]$ **do**
    $order.append([d-s, s])$
  end
end
return *order*

Finally, we provide a simple edit-history recovery method which operates on the values in M, E, and F in order to recover an edit transcript. The transcript is recovered from a traceback of optimal values in the array indices starting from (n, m) and going back to $(0, 0)$. The steps are sketched in Algorithm 8.4.

ALGORITHM 4 TRACEBACK

Data: Given arrays M, E, F, P_M, P_E, and P_F.
Result: Traceback for the optimal solution given as an edit transcript path listed as coordinates.
$v = max(M[m, n],\ E[m, n], F[m, n])$
$path = []$
$i, j = (n, m)$
$path.append((i, j))$
while $(i, j) \neq (0, 0)$ **do**

$v = \max(M[i, j], E[i, j], F[i, j])$

$$i, j \leftarrow \begin{cases} (i-1, j-1) \text{ if } v = M[i, j] \\ (i, j-1) \text{ if } v = E[ij] \\ (i-1, j) \text{ if } v = F[i, j] \end{cases}$$

$path.append((i, j))$

end

8.3.3 EXAMPLE ALIGNMENTS

To normalize two functions for comparison, we define a tokenization technique. Each token is compared to dictionary keys which map to integer values when they have been previously observed. If they have not been observed, new integers distinct from all the other values of the dictionary are assigned to them (Table 8.7).

The normalized view of our functions are ISPUNCT represented by the following sequence: 0 1 2 3 1 13 120 101 121 15667 26 101 8 2 15668 52 31 9 71 31 8 110 21 111 1017 21 112 90 21 44 1 45 9 4583 10 29 97 22 21 21 98 66 99 66 21 44 1 45 46 47 8 8; and ISSPACE represented by the following sequence: 0 1 2 3 1 13 120 101 121 242 26 101 8 2 243 52 31 9 71 31 8 110 21 111 244 21 112 90 21 44 1 45 9 245 10 29 97 22 21 21 98 66 99 66 21 44 1 45 46 47 8 8.

Both sequences have length 52.

The global alignment can be represented by the view

```
!"#$".;(<';()#aU@*h@)162n63{6M"N*h+>$766%c&c6M"NOP))
|||||||||X||||X|||||||||X|||||||||X|||||||||||||||||
!"#$".;(<W;()#XU@*h@)162Y63{6M"N*Z+>$766%c&c6M"NOP))
```

In this edit view, we can see the symbols which align with a pipe symbol linking them (ie, "—"), when mutations are introduced an "X" represents the edit which took place.

We include a few additional function comparisons which show a more interesting ability to create and extend a gap in order to accommodate additional matching later on in the string. Take, for example, the two functions LOOKUP and REMOVE-SYMBOL with the following alignment:

Table 8.7 Two Functions and Their Dictionary-Normalized Integer Representations

ISPUNC	Normalized	ISSPACE	Normalized
pushq %rbp	0 1	pushq %rbp	0 1
movq %rsp, %rbp	2 3 1	movq %rsp, %rbp	2 3 1
cmpl 27, %edi	13 120 101	cmpl 27, %edi	13 120 101
ja 0x100021f0f	121 15667	ja 0x100021bcf	121 242
movslq %edi, %rax	26 101 8	movslq %edi, %rax	26 101 8
movq 229669(%rip), %rcx	2 15668 52 31	movq 230501(%rip), %rcx	2 243 52 31
movl 60(%rcx,%rax,4),%eax	9 71 31 8 110 21	movl 60(%rcx,%rax,4), %eax	9 71 31 8 110 21
shrl 3, %eax	111 1017 21	shrl 4, %eax	111 244 21
andl, %eax	112 90 21	andl , %eax	112 90 21
popq %rbp	44 1	popq %rbp	44 1
ret	45	ret	45
movl 192, %esi	9 4583 10	movl 6384, %esi	9 245 10
callq 0x10004bd7a	29 97	callq 0x10004bd7a	29 97
testl %eax, %eax	22 21 21	testl %eax, %eax	22 21 21
setne %al	98 66	setne %al	98 66
movzbl %al, %eax	99 66 21	movzbl %al, %eax	99 66 21
popq %rbp	44 1	popq %rbp	44 1
ret	45	ret	45
nopw %cs:(%rax,%rax)	46 47 8 8	nopw %cs:(%rax,%rax)	46 47 8 8

```
!"#$"!&!'!R!(#-&>Q#)RJNNS]U)#&-#R=*$xJpp>)#)(<,(K^U(p_HKL(#((A((B',RC
 (Za#:(-#&=#
|||||||||||||||||X||||    X|X|||X||X|||||||||||||X||X|X|X|            XXX|X
 XX|X||||||X||
!"#$"!&!'!R!(#-'>Q#)---&SvU)#'-#&=*$xJpp>)#)R<,RKwURp----------x:),C
 (&Zy#:(-#'=#

RW>v766--------------Z---------a#r(---'#')M(MRM'M&M"NOP))
X||||||              |         X|X|    X|X||||||||||||| ||
&W>v766BzKL(#(R#R(A((Z{J,,p|R}U#H()#)R#()M(MRM'M&M"NO-))
```

This alignment, represented in two alignment rows of 80 symbols each, reveals that the algorithm with $\delta, \epsilon = -10, -1$, and $S(a, b) = \begin{cases} 10 \text{if} a = b \\ 0 \text{o.w.} \end{cases}$, can introduce sizable gaps (on both sides) in order to preserve string matching such as what we see early on in the head, as well as the tail.

When we reconsider the alignment with $\delta, \epsilon = -50, -10$, we get a slightly different answer which is likely to use mutation:

```
!"#$"!&!'!R!(#-&>Q#)RJNNS]U)#&-#R=*$xJpp>)#)(<,(K^U(p---------------
 _HKL(#((A((B
|||||||||||||||X||||    X|X|||X||X|||||||||||X||X|X|X|
 XXXXXXXXXXXX
!"#$"!&!'!R!(#-'>Q#)---&SvU)#'-#&=*$xJpp>)#)R<,RKwURpx:),C(&Zy#:(-#'
 =#&W>v766BzK
```

```
',RC(Za#:(-#&=#RW>v766Za#r('#')M(MRM'M&M"NOP))
XXXXXXXXX|XXXXXXXXXXXXXXXXXX|X|||||||||||||  ||
L(#(R#R(A((Z{J,,p|R}U#H()#)R#()M(MRM'M&M"NO-))
```

This simple example with various affine alignment parameters shows that the method is flexible and able to accommodate a large number of similarity concepts. In particular, it forms a demonstration that alignment algorithms are valuable computational tools capable of identifying various similarity notions in comparable objects.

8.4 SUMMARY

Strings are common in cybersecurity data. String data in cybersecurity problems are usually formal encodings of data, objects, or functions. Due to large volumes of data, efficient techniques for analyzing strings are needed. Cryptographic and universal hash functions have also been utilized for their efficiencies; however, it's important to understand both their properties and limitations concerning the possibility of hash collisions. We present several examples of how hash functions may be used to find common objects with the bag of numbers technique. The bag of numbers technique is extended with the normalized bag of numbers to retain the efficiencies and precision of hashing but address the effects of noise patterns in data.

In the context of reverse engineering and code analysis, where little or nothing is known about the code's provenance, the alignment algorithms offer distinct value in finding pattern motifs, such as calling conventions, or in identifying commonly occurring structures. In the design of protocols, we often hear the analogy that signals can be constructed like trains (a sequence of engines and cars comprise a train), and we may expect common patterns and sequences in particular in the header and trailer sections. So, perhaps one way to consider the utility of alignment algorithms is that it can do well in matching such preserved patterns and identifying the variants. The Needleman-Wunsch algorithm can be used to determine motifs and sources of variations even when we know little about its origin or history.

CHAPTER

Persistent homology

9

Statistics teaches us how to ask predictive questions about the data set we analyze. For example, we can ask the question "does this new point fit in with our current data set?" Instead of asking predictive questions, we could ask "what is the structure of our data set" and determine features of it.

The point of this chapter is to introduce you to the mathematical fields of homology and persistent homology and how these fields are used to understand the structure of data sets. There are many algorithms to compute Persistent Homology. In depth knowledge of these algorithms is not the focus, rather, we are interested in understanding the results they yield.

Persistent homology is a relatively new field for data analysis. It is being slowly applied to cybersecurity problems, for example, in analyzing wireless traces. It has also been applied to analyzing traffic patterns by examining how much traffic traverses a network.

To begin, consider A **point cloud** which is some set of data points in $\mathbb{R}^n$. All of the points in the set are collected in a similar fashion. For example, medical scanning devices create a point cloud of the human body in $\mathbb{R}^3$. Another method to create a point cloud is to follow a Roomba as it moves through a room. We collect the position of the Roomba every n seconds, which allows us to study the path the Roomba followed. A third example of a point cloud is the pixels colored black when an **X** is displayed on a computer screen.

An example of a point cloud is in Fig. 9.1.

We can see from the shape of the point cloud that this is the letter **B**. That requires human eyes to tell what the data looks like and it is aided by the fact that a letter **B** has two distinct holes. Suppose we are looking at a photo of a **B** that has been distorted by distance or attempts to enhance it. Noise, meaning extra data points, often appears in such attempts. So we are not really sure if we are looking at a **B** or perhaps a **O** or **D** that has extra smudges. It is even possible that it is an **E**. We need a way to find and quantify the holes in the data structure without relying only on our eyes. This arises in handwriting recognition as well. Everyone has a slightly different way of writing a **B**, so the ability to quantify exactly what makes a **B** is important.

Homology and persistent homology are very focused on finding holes in our data sets. For example, suppose we have a field that is covered by sensors. The ability to

Cybersecurity and Applied Mathematics. http://dx.doi.org/10.1016/B978-0-12-804452-0.00009-9

FIG. 9.1

A point cloud.

find holes in this sensor net is very important for both the defender and the attacker. The defender would like to remove the holes, while the attacker would like to take advantage of them.

The methods discussed in this chapter work best with data in either $\mathbb{R}^2$ or $\mathbb{R}^3$. For ease of explanation and visualization, we will assume all of our data sets in this chapter are in the plane, that is, $\mathbb{R}^2$. In other words, we are looking at two-dimensional shapes, like circles, squares, rings, and letters.

9.1 TRIANGULATIONS

A point cloud is generally unordered. There is no defined starting point or ending point, as you would find in a connect-the-dots picture. We can look at Fig. 9.1 and generally get an idea of what points are related to each other, but if our point cloud is much more amorphous, then at first glance, we would not be able to determine a relationship.

We can try to create an analyze a graph of the points to consider the shape of the data. In particular, consider the case where the graph is a polygon. For our example point cloud in Fig. 9.1, Fig. 9.2 is a polygon that preserves the structure along with displaying the **B**. On the other hand, Fig. 9.3 is also a polygon that preserves the structure of the point cloud. Since we know that Fig. 9.1 is a **B**, then obviously Fig. 9.2 is a polygon that correctly represents the shape of the **B**, but that is based

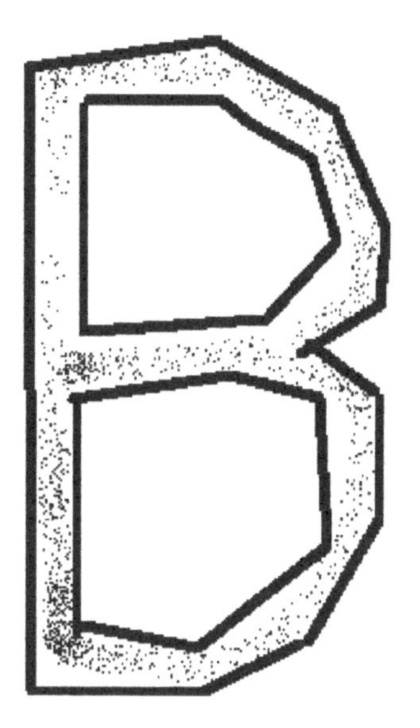

FIG. 9.2

A polytope for **B**.

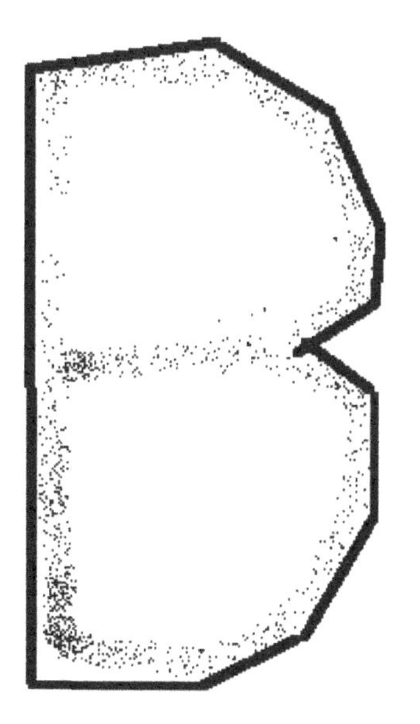

FIG. 9.3

Another polytope for **B**.

on foreknowledge of the point cloud. We cannot expect to have a priori knowledge of our sets.

The smallest convex polygon that encloses the point cloud is called the **convex hull**. For the letter **B** it would look something like Fig. 9.3. This is because a convex set means that every two points in the region are connected by a straight line. The problem with the convex hull is that we have lost our holes again since it is a convex set. A hole in the polygon means that there are at least two points that cannot be connected by a straight line because the straight line between them goes through the hole.

Returning to our letter **B**, Fig. 9.2 is still the best polygon for the point cloud. The convex hull is not the best polygon for the shape of the letter **B**, but it does provide an outside limit for how big the polygon can be. It is possible to use a larger polygon, but we want to use the closest fit to our data. If we use a polygon that is too large, then the structure we create can overwhelm the structure of the points.

Let us consider the points themselves as a set $P = \{p_1, p_2, \ldots, p_n\}$. We can partition the plane containing the points into smaller convex hulls, where each of the smaller convex hulls has only one point from the set. This gives us a set of n convex hulls and we will call them $\{C_1, C_2, \ldots, C_n\}$, one for each p_i. Each C_i is then all of the points q where $d(p_i, q) < d(p_j, q)$ for all $p_j \in P$ and $p_j \neq p_i$. In other words, each convex hull is the set of points that are closer to the center of the hull than to any other point.

This method partitions the plane into what we will call **areas of influence** for each point in the point cloud. The diagram itself is actually called the **Voronoi diagram**. Fig. 9.4 is a Voronoi diagram for a set of points in a ring shape. You can see how it separates the plane into independent regions that share borders. The regions within the borders are the points closest to the center, so this is an effective partition of the plane into distinct areas. Each distinct area is also called a **Voronoi cell**.

The Voronoi diagram is actually quite useful. Suppose we have 10 fire stations in a region. The Voronoi diagram for the fire stations would be computed in that region so we can determine the closest fire station for any house in the region. This is called the **nearest neighbor problem** and the Voronoi diagram is a solution for it. An early application of the Voronoi diagram was in tracking the origin of a cholera outbreak. The doctor was able to use a diagram to trace the outbreak back to a particular water source. A third application is related to wireless networking. If we build a Voronoi diagram of the wireless network using the set of access points as our point cloud, then the resulting diagram tells us how much space each access point is expected to cover on its own. If we know the range of the wireless access points, then we also can figure out how well our area is covered. It does not highlight the holes in the coverage though. It merely reveals the area served by each access point.

There is no requirement that the distance used in the creation of the Voronoi diagram be the Euclidean distance. The Manhattan distance will also create a diagram, as will the Minkowski metric. Since we are working with $\mathbb{R}^2$ in this chapter, any metric defined on $\mathbb{R}^2$ will work.

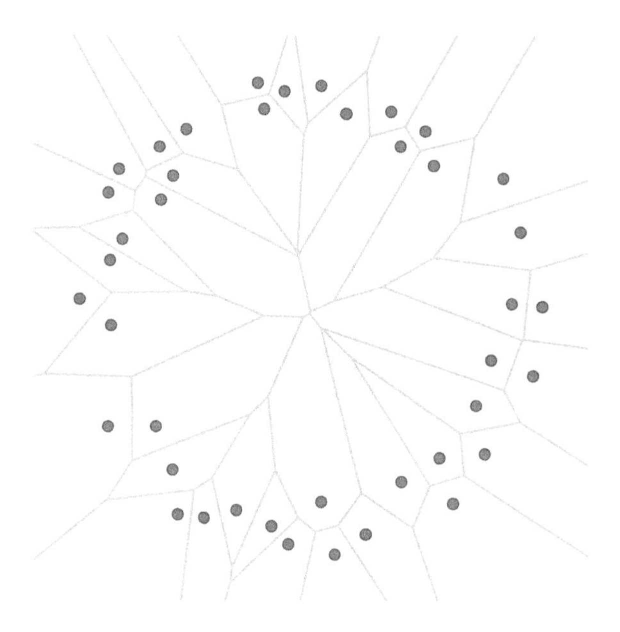

FIG. 9.4

A Voronoi diagram.

We can use the Voronoi diagram to create relationships between the points in the point cloud. We will say two points in the set P have an edge between them if their Voronoi cells share a border. Fig. 9.5 is a graph on the same set of points as in Fig. 9.4. The graph is made up of triangles, and in fact it can be proven that any graph created from the Voronoi diagram in this manner is made of triangles. It is called the **Delaunay triangulation**.

There is another method that creates the Delaunay triangulation. In this case, it is created directly from triangles formed in the point cloud. Clearly, any three points in the point cloud defines a triangle. We do not want to use all of the triangles as we need a specified relationship between the edges in the triangle to add it to the triangulation. To accomplish this, we define a **circumcircle** of a triangle. It is a circle that passes through all three vertices. Given all of the possible triangles in the point cloud, we will only chose the triangles where the circumcircle of the triangle contains no other points in the point cloud.

Delaunay triangulation also maximizes the minimum angle in the triangles. That is, if we find another triangulation that creates triangles with a larger minimum angle, then we can prove that that triangulation is a Delaunay triangulation. This is useful when the triangulations are used to model terrain. It can also be used in morphing, that is when one picture is distorted so that it looks like another.

Suppose we have two triangles that share an edge. Then we can erase the shared boundary and create a shape with four sides. We can repeatedly do this with the

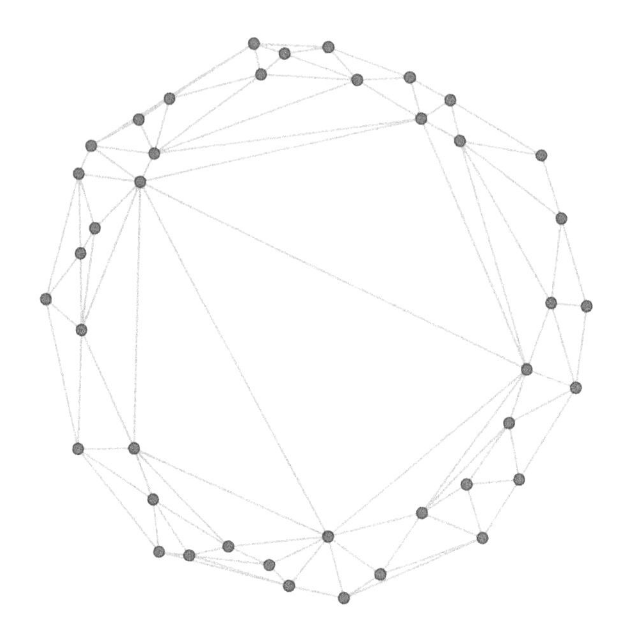

FIG. 9.5

A Delaunay triangulation.

triangles in the Delaunay triangulation and what we end up with is the convex hull. This tells us that the Delaunay triangulation is defined in such a way that it preserves the generic shape of the points, while it does not find our holes.

From Fig. 9.5 we see we have connected the points in the original point set in a reasonable way, but we are still obscuring the hole that is present in the data. We still need a method that will allow us to find the holes, but by creating a graph on the point we have made an important first step.

9.2 α SHAPES

Returning to Fig. 9.5, the Delaunay triangulation of a ring, we can see that if we ignore the biggest triangles we have connected the dots to form a graph of a ring. Figuring out which triangles to remove means we know something about our data set to begin with. We can tell that the larger triangles obscure the holes, but how we determine what "larger" means. Also, there may be smaller holes of interest in our point cloud. Removing the large triangles reveals large holes but leaves smaller holes obscured. If we are modeling a sensor network, then the smaller hole still leaves an area vulnerable for an attacker.

In the second definition of the Delaunay triangulation, we use a circumcircle to determine if a trio of points is in the triangulation. Three points are in the triangulation if their circumcircle contains no other points in the point cloud which gives us a set of circles along with the triangles in the triangulation. For each triangle in the Delaunay triangulation, we will let σ_i be the circumcircle of it. This gives us the set $\Sigma = \{\sigma_i\}$.

We can create subsets of Σ based on the radius of the circumcircle. For a value $\alpha > 0$, we will define the set as Σ_α. This will contain all of the $\sigma_i \in \Sigma$ where the radius of the circumcircle is less than or equal to α. Suppose $\alpha_a \leq \alpha_b$. Since Σ_{α_b} contains all of the triangles where the radius of σ_i is less than α_b, then every triangle in Σ_{α_a} is in Σ_{α_b}. In other words, $\Sigma_{\alpha_a} \subseteq \Sigma_{\alpha_b}$. The triangulation created by doing this is called the α **shape**.

As we said earlier, we can use the Delaunay triangulation to create the convex hull by removing shared edges. If we remove shared edges of the α shape, we create what is known as the α **hull**. An α hull for our ring example is in Fig. 9.6.

The value of α chosen for this picture has maintained the hole in the ring, which is what we were looking for. On the other hand, a smaller value of alpha might show more than one hole, which is not what we are looking for. It would be useful if there as an exact value of α that would illuminate the holes in the data. Fig. 9.6 is an example of a single value that accomplishes this. However, in this case we know that there is one hole in the point cloud and that it is one size. This does not apply when we have data that is more varied. If we have multiple holes of different sizes, one value may expose some holes while it obscures others.

There are a couple of special cases for α. If $\alpha = 0$ then the radius of the circle is 0, so we consider that as the set of points in the point cloud. There are no triangles in this case. We also know from the definition of the α shape that there is a value of α where the α shape is the Delaunay triangulation and the α hull is the convex hull. If we increase α after we have reached that point, then the shape of the data does not change.

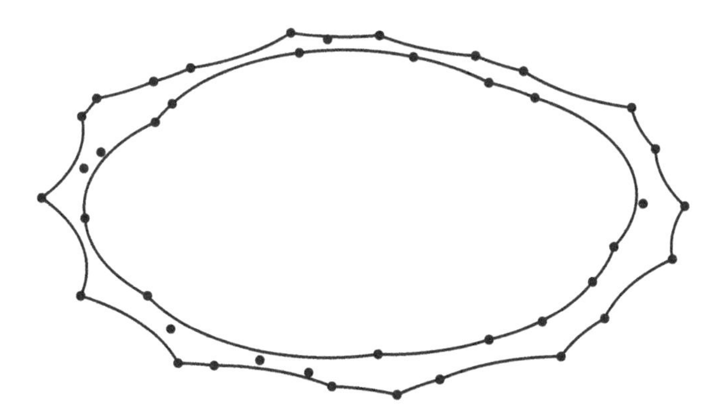

FIG. 9.6

An α hull.

9.3 HOLES

We are going to digress a bit and discuss the nature of holes in data. We start by defining a **region** or a subset of $\mathbb{R}^2$. For example, a square, a circle, several circles, a ring, a square with several rings inside of it, a triangle, or even a hexagon with three triangles in it are all regions.

A **continuous deformation** of a region is where we can shrink or twist the region but without tearing. Gluing is also not allowed. This allows us to transform one region into another. In other words, it is terraforming, or earth shaping, with restrictions. We can expand our land, shrink our land, or even change the borders completely. We also can expand a water feature or shrink it, but we cannot remove it entirely. Removing it is essentially gluing a patch on the region, which is not allowed. We cannot split the land into two pieces or take two pieces of land and put them together to form one.

For example, suppose we have a circle in $\mathbb{R}^2$. We can shrink it or expand it or even push it until it looks like a square. We cannot rip it open and create a line out of it though. Think of it as if the region was enclosed with elastic strings. We can shrink the string or expand it, but we cannot tear it. So if we had a ring with two strings, we could push the inner string out so that it matches the outer string, but we cannot remove the strings. Similarly, suppose we have two separate circles. There is no way to join them into one circle without tearing them open.

Let us look at another example. Consider the letter **A**. We can push the legs up until they go away, which leaves us with a triangle. We can then deform the triangle so that it looks like the letter **O**. Conversely, we can deform the letter **O** so that it looks like a small triangle then pull on the triangle to form legs. On the other hand, we cannot continuously deform an **A** to form a **H**. We would have to tear the triangle apart to create the **H**.

We can also look at this in terms of graphs. The holes in a graph are the fundamental cycles. If an edge is not part of a fundamental cycle, then it is not part of any cycle in the graph. This is because any cycle in the graph can be created as a linear combination of the fundamental cycles in the graph. So we can push the edge that is not part of any cycle in the graph back into the graph by using a continuous deformation. In a similar fashion, two components of a graph cannot be joined together using a continuous deformation to create a new graph.

In summary, a continuous deformation cannot remove a hole in a region, nor can it put two pieces of a region back together. This means that we can classify a region by the number of holes that it has. A region with two holes can be deformed into a similar region with two holes, but cannot be deformed into a region with three holes. Similarly, if a region has two distinct parts, we cannot combine them to form one nor rip one apart to form three.

We have two kinds of holes possible in a region of $\mathbb{R}^2$. The first is the number of distinct subregions in the region. These are separate in that we cannot glue them together to form a new region. For example, the letters **o o** are two distinct regions that cannot be continuously deformed to form a new region. For example, we cannot put them together to form a letter **B**. The second type of hole is the region that cannot be removed without tearing. For example, the letter **o** cannot be continuously deformed to remove the hole. These holes are the features of the region, also known as the **topological features**.

9.4 HOMOLOGY

As we said in the previous section, the holes in a region can be used to classify it. One way to study this is called **Homology**. Homology gives us a rigorous method to define exactly what a hole is. We can see it with our eyes, but homology gives us the math. As in the previous section, we have two types of hole. Each level of hole is called a **dimension**. The zero-dimension is the number of pieces the region has and the one-dimension is the number of holes as we think of holes in $\mathbb{R}^2$. We compute Homology on a per dimension basis, so there is one value for dimension 0, one for dimension 1, and so on.

Homology is defined on a region called a **simplicial complex** or just a **complex**. A simplicial complex is a region that is built by gluing smaller regions together. These smaller regions are called **simplices** (or simplexes) in the plural form. A single region is called a **simplex**. We build a larger simplex from a smaller simplex by adding a point and connecting every point in the smaller simplex to the new point. The smallest simplex is a single point. We add a point, connect it to the first point with a line, and we have a 1-simplex. Repeating the process yields a triangle, the largest simplex in $\mathbb{R}^2$. If we add another point, we create a tetrahedron, which is a simplex in $\mathbb{R}^3$. Each simplex has a dimension. The point is a zero-dimensional simplex, a line is a one-dimensional simplex and a triangle is a two-dimensional simplex.

Each simplex, other than the zero-dimensional simplex, has a **face**. The face of a n dimensional simplex is the $n - 1$ dimensional simplexes contained within it. For a triangle, the faces are the three line segments that create the triangle. For a line segment, the faces are the points at each end. A simplicial complex is the union of all of the simplexes, including the faces of the simplexes. This means we cannot have a triangle in the simplicial complex without having the edges that make up the triangle in the complex as well.

Homology is associated with a field of mathematics called Group Theory, a beautiful but complex subject. In essence, group theory gives us a way to abstract properties of sets that have an operation associated with it. Such sets are called **groups**. For example, we can associate the operation addition with the set of integers. Not every set with an operation is a group. If we do have a group, one property is that it may have **generators**. A generator is an element of a group that can be used to produce, or generate, every other element of the group. An important property of a group is how many generators there are in a given group.

The homology of a simplicial complex at each dimension is a group called the **homology group**. The generators of the group correspond to holes in the complex at that dimension. In order to know the number of holes in a simplicial complex at each dimension, we need to know the number of generators in the homology group at each dimension. The number of generators are known as **Betti numbers** and are denoted by β_i where i is the dimension of homology we are computing. For $\mathbb{R}^2$, we only have β_0 and β_1 to compute.

We know that the circle is a single element, in that everything is connected. So there is one generator in dimension 0. The circle also has a single hole, so there is one generator in dimension 1. So $\beta_0 = 1$ and $\beta_1 = 1$. If the Betti number of a region is 2, then there are two topological features at the given dimension. For example, if we have two disks in the plane, then there are two holes at dimension 0 so the Betti number is 2. Each of the disks has their own generator.

There are many algorithms to compute homology of a simplicial complex and to find the number of generators in each dimension. We are not concerned with these algorithms, but it is important to be able to understand these algorithms tell us. They give us a well defined method of finding the number of holes in a simplicial complex. They are also not affected by a continuous deformation of a region. A square will have the same number of generators as a circle and the letter **B** will have the same number of generators as the number **8**.

9.5 PERSISTENT HOMOLOGY

Homology tells us the number of holes in a system and α shapes allow us to find a hull that will allow us to determine the number of holes in data. We are going to marry the two to create **Persistent Homology**.

For the example in Fig. 9.6, we know that there is one hole in dimension 0 and one in dimension 1 and if we find the value of α that allows us to compute that via homology, we can show that. Except that requires that we know that there is one hole in dimension 0 and one hole in dimension 1.

In general, we do not know the structure of the data and we want to determine what it is. This mean that there is no definitive value of α that will show us the structure unless we already know what we are looking for. On the other hand, we know that if we have a sequence of increasing values of α, call them $\alpha_0 \leq \alpha_1 \leq \cdots \leq \alpha_n$ then we can create the series of subsets $\Sigma_{\alpha_0} \subseteq \Sigma_{\alpha_1} \subseteq \cdots \subseteq \Sigma_{\alpha_n}$. We can compute the homology of the α hull at each state and look for the holes that persist over each value of α. In other words, we are using persistent homology to find persistent holes. We do this by finding the generators of the homology group in dimensions 0 and 1 at each α_i. At α_{i+1}, some of these generators may remain and some may be removed as the hole they represent is obscured.

When $\alpha = 0$, then we have just the point cloud. If there are n points in the point cloud, we have n zero-dimensional holes. As α grows, some of these points get absorbed into other holes and we have fewer and fewer until α grows large enough that we have the convex hull. When α grows large enough, we will have only one hole of dimension 0 and no holes in dimension 1. This is true of the convex hull of any point cloud.

For every α in the sequence $\alpha_0 \leq \alpha_1 \leq \cdots \alpha_n$ we compute the Homology group in both dimensions. Thus, we have a value of α such that a generator is created and a value of α for which it no longer exists. This gives us an interval for which a given generator exists. Each point in the point cloud has an interval associated with it, at least for dimension 0.

The first value of α for which a hole or topological feature exists is called the **birth**. The last value for which it exists is called the **death**. That is to say, the first time the feature exists is the birth of the generator and when it no longer exists is the death. Generally features do not just disappear. They get subsumed into another feature as α grows. A one-dimensional hole might disappear when α grows large enough to obscure it.

The result of persistent homology is a collection of intervals. Each interval consists of the birth of a topological feature and the death of the feature. If the birth equals the death, then the feature appeared at the value of α and disappeared immediately. It is also important to note that a feature cannot die before it is born. In this way we can identify the lifetime of features in the point cloud. The more persistent the feature, the more likely it is represented by the points in the cloud.

9.6 VISUALIZING PERSISTENT HOMOLOGY

We can visualize this by using a method called a **barcode**. It is a collection of the intervals found in persistent homology and displayed as such. The x-axis is the value of α and the y-axis is this the homology generator. The homology generators are generally unordered. Each interval found in persistent homology is drawn on the chart as a line, starting at birth of the topological feature represented by the generator and ending at death.

The features that are born and die quickly are generally left off of the bar chart because otherwise we have a lot of noise that clutters the chart. We really want to find the features that persist, because they are the structure that we are looking for. So for some value of δ that is chosen based on the data, we ignore features that live for less than δ. In other words, we ignore features where $\alpha_{\text{death}} - \alpha_{\text{birth}} <= \delta$.

Fig. 9.7 is a point cloud in the shape of a circle. There are obvious gaps in the point cloud, but it still basically has a circular shape. We know that at some value of α, the hole in the middle of the circle will be obscured. The barcode for this circle is in Fig. 9.8.

The two long lines represent a persistent hole in both dimension zero and one. The longer line is for dimension zero, since it remains persistent through the point where all other holes disappear. The shorter line is the hole for dimension one, since at some value of α the hole disappears. There are also several short lines. This is due to the holes in the point cloud, causing gaps in the outline of the circle. Once α grows large enough, those holes are absorbed into the larger zero dimensional hole.

Another method to compute and visualize persistent homology is by using the **persistence diagram**. This does not use the α shapes, rather it is a more discrete

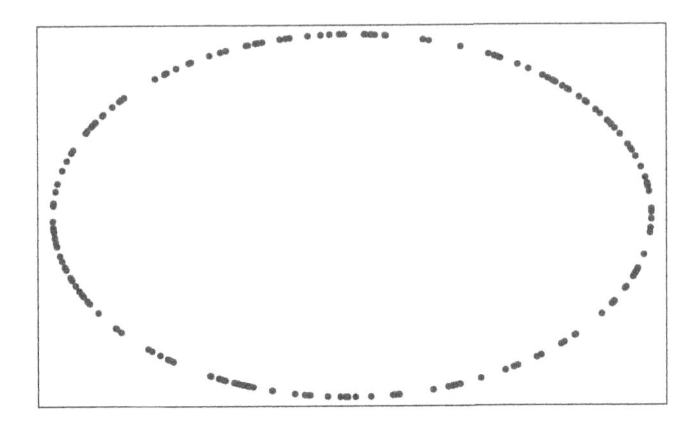

FIG. 9.7

A circular point cloud.

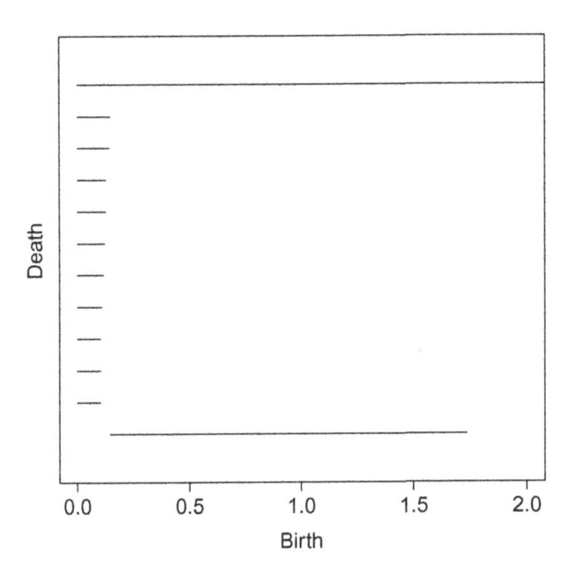

FIG. 9.8

The barcode for Fig. 9.7.

case. We start with a simplicial complex and add another simplicial complex to create a region. The second may be glued to the first or it may be a disjoint region. We do know that by using this method to create a new region, the first region is contained in the second. For a simpler case, start with a point and glue simplices to the first point, repeating the action as many times as desired.

In other words, we are building a series of sets where $\emptyset = K_0 \subseteq K_1 \subseteq K_2 \subseteq \cdots \subseteq K_n$. Imagine it as building the Delaunay triangulation triangle by triangle. We start by picking a single triangle out of the triangulation as K_1. Then we add another triangle from the triangulation to create K_2, another to create K_3, and we continue until we have created the Delaunay triangulation. At each i we can create the Betti number of the complex for dimension d. We will call this β_d^i.

We can also compute the number of topological features or holes that appear at K_i and disappear at K_j for dimension d. For example, suppose we have a hole that appears at K_i but the addition of two triangles obscures the hole at K_{i+2}. We denote the number of topological features that are born at i and disappear at j in dimension d as $\beta_d^{i,j}$. We can now compute multiplicities of the Betti number in Eq. (9.1):

$$\mu_d^{i,j} = \left(\beta_d^{i,j-1} - \beta_d^{i,j}\right) - \left(\beta_d^{i-1,j-1} - \beta_d^{i-1,j}\right) \tag{9.1}$$

The first difference, $\beta_d^{i,j-1} - \beta_d^{i,j}$, counts the topological features in K_{j-1} that are born at K_i and die at or before K_j. In short, this is the number of features in K_{j-1} that die before K_j. The second difference, $\beta_d^{i-1,j-1} - \beta_d^{i-1,j}$, counts the number of topological features that are in K_{j-1} that are born at or before K_{i-1} that die before K_j. So the difference must be the number of topological features that are born in K_i that die in K_j.

We define the persistence diagram as the collection of points (i, j) where $\mu_d^{i,j} = 1$. The first coordinate is the birth of the topological feature and the second is the death of it. We can plot this on the plane where the x-axis is the birth and the y-axis is the death of the topological features. The values of the persistence diagram are above the line where $x = y$ in the plane. This is because for values below the diagonal, death is less than birth, which cannot occur. There are none on the diagonal because that would imply that birth equals death, something we do not measure.

There is another definition of the persistence diagram that is more in line with the barcodes, making it easier to compute. Using the sequence of α, $\alpha_1 \leq \alpha_2 \leq \cdots \leq \alpha_n$, we plot $(\alpha_{\text{birth}}, \alpha_{\text{death}})$ for each generator, where as in the previous definition of the persistence diagram, the x-axis is the birth values and the y-axis is the death values. In both cases we are plotting the same essential values, the birth and death of topological features. The first case uses a discrete method to build the simplicial complex from simplices, the second uses the α shapes.

Returning to the circle in Fig. 9.7, we can use the second definition of persistence diagram, which gives us Fig. 9.9.

In this case, we have a cluster of points near zero. These represent the same features as the short lines in Fig. 9.8. Since they are so close together, we do not know how many of them there are. The square in the diagram represents the birth and death of the topological feature in dimension one.

So the barcode makes it easier to see and count those smaller topological features that are born and die at close to the same value of α, but the persistence diagram

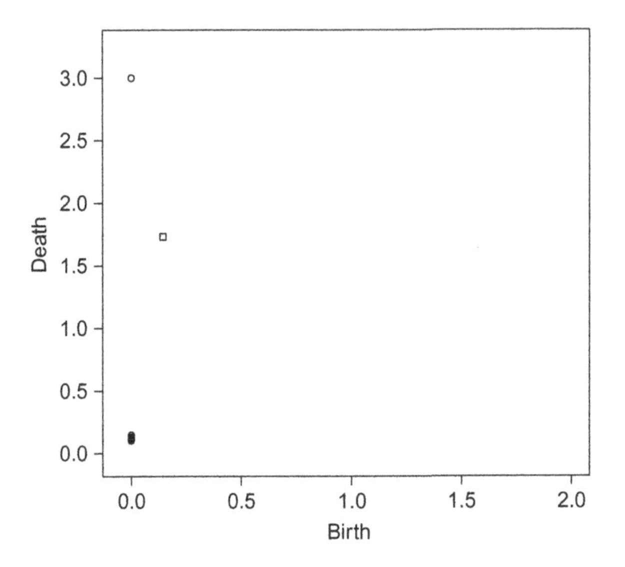

FIG. 9.9

The persistence diagram for Fig. 9.7.

allows us to easily label the topological features in each dimension by using different point types. Each method to display the persistent homology of a point cloud is valid, but depending on the data, one may be easier to read than the other.

9.6.1 COMPARING POINT CLOUDS

Aside from finding the structure represented by a point cloud, persistent homology can be used to compare the structures found within point clouds. In Section 9.3 we said that the holes in a region distinguish it. SO it is logical to assume that the holes we find in persistent homology distinguish a point cloud. In order to accomplish this, we will need a metric that will allow us to determine the distance between the persistent homology of point clouds. We accomplish this by comparing the persistent diagrams of the clouds.

The simplest metric to find the distance between two persistence diagrams is the Hausdorff metric. Let $\mathcal{P}_1$ and $\mathcal{P}_2$ be two persistence diagrams. The Hausdorff metric relies upon another metric in order to compute the distance. We will use the L^∞ distance. Recall that this is the maximum of the absolute value difference of the coordinates in the vector. The Hausdorff metric is in Eq. (9.2):

$$d_H(\mathcal{P}_1, \mathcal{P}_2) = \min_{p_1 \in \mathcal{P}_1, p_2 \in \mathcal{P}_2} d_\infty(p_1, p_2) \tag{9.2}$$

A more precise metric is the **Wasserstein** metric. It finds the best distance by attempting to match the two diagrams and computing the distance of the matchings. For the Wasserstein metric, we use all of the points in the persistence diagram, not just the ones we plot. This means that all intervals where birth equals death are used.

Let $\eta : \mathcal{P}_1 \to \mathcal{P}_2$ be a bijection between persistence diagrams $\mathcal{P}_1$ and $\mathcal{P}_2$. Then for every $p \in \mathcal{P}_1$ there is one and only one $\eta(p) \in \mathcal{P}_2$. We also know that for every element $q \in \mathcal{P}_2$ there is an element $p \in \mathcal{P}_1$ such that $\eta(p) = q$.

Now for every $p \in \mathcal{P}_1$, we will compute the L^∞ metric between p and $\eta(p)$. This gives us a set of distances between the two sets, so we can find the maximum of that set. Now we take every possible bijection between $\mathcal{P}_1$ and $\mathcal{P}_2$ and compute the maximum for each one of those. That gives us a second set. The Wasserstein metric is then the minimum of the distances found for all of the bijections. Eq. (9.3) is a precise definition of the metric:

$$d_W(\mathcal{P}_1, \mathcal{P}_2) = \min_{\eta} \max_{p \in \mathcal{P}_1} d_\infty(p, \eta(p)) \tag{9.3}$$

The Wasserstein metric is bounded by the Hausdorff metric. This means that $d_W(\mathcal{P}_1, \mathcal{P}_2) \leq d_H(\mathcal{P}_1, \mathcal{P}_2)$.

Using the Wasserstein metric allows us to determine the distance between the persistence diagrams, and in turn, the distance between the structures found within the point clouds. For example, it can be used to determine if a point cloud is more like a **O** or a **B**. This is accomplished by computing the persistent diagrams of a **O**, a **B** and the point cloud in question, and then applying the Wasserstein metric.

9.7 CONCLUSIONS

In this chapter we have discussed how the structure of point clouds can be examined using persistent homology. We have discussed point clouds only in the plane, but it can also be extended to $\mathbb{R}^3$. In $\mathbb{R}^3$, a hole becomes a void. Think of it as the interior of a ball. We can shrink the ball or expand it, but we cannot remove the void without ripping the ball open. The Delaunay triangulation, Voronoi diagram and α shapes can all be constructed in $\mathbb{R}^3$. In fact, the Delaunay triangulation is often used in modeling terrain.

APPENDIX

Introduction to linear algebra

A vector is an element of $\mathbb{R}^n$ where $n \geq 1$. At its core, linear algebra is the study of sets of the vectors and functions between the sets. It also includes **matrices**, a two-dimensional arrays of numbers. Matrices and vectors can be used in graph theory, game theory, text mining, encryption and more. In this appendix we cover the basics of linear algebra and include properties useful in this book.

A.1 VECTOR ALGEBRA

A **vector space** is an ordered set of vectors. We write a vector as Eq. (A.1) and we can denote a vector space as V. A vector space is a subset of $\mathbb{R}^n$. This means that every vector in the vector space has the same number of elements:

$$\bar{v} = [v_1, v_2, \ldots, v_k] \tag{A.1}$$

We say that $\bar{v} \in \mathbb{R}^k$ to indicate that it is as a vector in a vector space $\mathbb{R}^k$.

A.1.1 VECTOR ADDITION

If $\bar{v}$ and $\bar{w}$ are vectors in $\mathbb{R}^n$, then we can add them. The addition is done on an element-by-element basis. v_i and w_i are added together for each $0 \leq i \leq i$. We do this as Eq. (A.2):

$$\bar{v} + \bar{w} = [v_1 + w_1, v_2 + w_2, \ldots, v_k + w_k] \tag{A.2}$$

We can also subtract two vectors, using the same method.

We cannot add or subtract two vectors where $\bar{v} \in \mathbb{R}^n$ and $\bar{w} \in \mathbb{R}^m$ and m is not equal to n. They have to both be in $\mathbb{R}^n$.

If we add a vector of all zeroes, that is, $\bar{0} = [0, 0, \ldots, 0]$ to another vector v then we still have v. Similarly as with addition for real numbers, given any vector $\bar{v} \in \mathbb{R}^n$ there is another vector $-\bar{v} \in \mathbb{R}^n$ such that $\bar{v} + -\bar{v} = 0$

A.1.2 VECTOR SCALAR MULTIPLICATION

A scalar is an element of $\mathbb{R}$. We multiply a vector by a scalar in an element-by-element basis as illustrated in Eq. (A.3) for a scalar μ and a vector $\bar{v}$:

$$\mu\bar{v} = [\mu v_0, \mu v_1, \ldots, \mu v_k] \tag{A.3}$$

A.1.3 LINEAR MAPPINGS

If we have two vector spaces U and V, then we can have a function $f : U \to V$ between them. If this function preserves vector addition and scalar multiplication, then it is a **linear mapping**. This means that for any two vectors $\bar{u}, \bar{v} \in U$, then $f(\bar{u}+\bar{v}) = f(\bar{u}) + f(\bar{v})$. Also, for any scalar $\mu \in \mathbb{R}$, then $f(\mu\bar{u}) = \mu f(\bar{u})$.

Example A.1.1. If U is a vector space, then the function $id : U \to U$ given by $id(\bar{u}) = \bar{u}$ is a linear mapping. It is a special function known as the **identity** map.

Example A.1.2. Let $\alpha, \beta, \gamma, \delta \in \mathbb{R}$. Let $U \subseteq \mathbb{R}^4$ and $\bar{u} \in \mathbb{R}^4$ be given by $\bar{u} = [u_1, u_2, u_3, u_4]$. We define a function $M : U \to U$ by $M(\bar{u}) = [\alpha u_1, \beta u_2, \gamma u_3, \delta u_4]$. This is a linear mapping.

Example A.1.3. Letting $k = 1$, then we consider the function $f : \mathbb{R} \to \mathbb{R}$ given by $f(\bar{v}) = \bar{v}^2$. In other words, we are mapping an element of $\mathbb{R}$ to the square of the element. This is not a linear mapping. If it were, then $(x+y)^2$ would equal $x^2 + y^2$, which is not true.

A.1.4 MATRICES

The term **matrix** derives from the Latin term for **mold** or form. We think of it as a two-dimensional array of values. Eq. (A.4) is an example of a matrix:

$$\begin{bmatrix} a_{1,1} & a_{1,2} & \cdots & a_{1,n} \\ a_{2,1} & a_{2,2} & \cdots & a_{2,n} \\ & & \cdots & \\ a_{m,1} & a_{m,2} & \cdots & a_{m,n} \end{bmatrix} \tag{A.4}$$

Matrices are organized into **rows** (horizontal) and **columns** (vertical). The size of a matrix is the number of rows and columns, referred to as a (row,column) matrix. For example, a 3,4 matrix has three rows and four columns. We refer to the ith, jth entry of A as $a_{i,j}$ referenced by finding the value in the ith row and jth column. A **square** matrix a matrix in which the number of rows equals the number of columns.

A vector, or row vector, is a matrix with 1 row and n columns, also known as a $(1, m)$ matrix. A column vector is a matrix with n rows and 1 column, also known as a $(m, 1)$ matrix.

The set of real-valued matrices is denoted by $\mathbb{R}^{(m,n)}$. For each matrix we can define a linear mappings from the vector space $\mathbb{R}^n$ (the column space) to $\mathbb{R}^m$ (the row space). For the matrix A in Eq. (A.4), we define the linear mapping coordinate by coordinate. Eq. (A.5) defines the ith coordinate for a vector $\bar{v} \in \mathbb{R}^n$:

$$w_i = \sum_{k=1}^{n} a_{i,k} v_k \tag{A.5}$$

In other words, we multiply each element in the ith row by the kth element of the vector and summing the results. This becomes the ith element of the new vector. We can also think of this as multiplying a matrix by a vector, that is, $A\bar{v} = \bar{w}$. We can show that this is a linear mapping.

We can also multiply a row vector by a column vector. For this to work, the number of columns in the matrix must equal the number of rows in the vector. Eq. (A.6) is an illustration of our column vector $w \in \mathbb{R}^n$:

$$\begin{bmatrix} w_{1,1} \\ w_{2,1} \\ \cdots \\ w_{n,1} \end{bmatrix} \tag{A.6}$$

Eq. (A.7) is the jth, 1 element of the vector $\bar{w}$ after it has been multiplied by A:

$$x_{j,1} = \sum_{i=1}^{n} a_{j,i} w_{i,1} \tag{A.7}$$

We multiply every element of the jth row with every element of the column vector and sum the results.

A.1.5 MATRIX ADDITION

If we have two matrices of the same size, we can add them as we add vectors. The addition is done on an element-by-element basis. To be precise, let A and B be (n, m) matrices and let $C = A + B$. $c_{i,j} = a_{i,j} + b_{i,j}$. Similarly, we can subtract two matrices.

Example A.1.4. We begin with two $(3, 3)$ matrices A and B in Eq. (A.8):

$$A = \begin{bmatrix} 0 & 1 & 3 \\ 1 & 0.5 & 0.5 \\ 2 & 1.5 & 0.5 \end{bmatrix} \quad B = \begin{bmatrix} 1 & 3.5 & 1 \\ 2 & 3 & 3.5 \\ 0 & 3 & 0 \end{bmatrix} \tag{A.8}$$

Then the sum of the two matrices is given in Eq. (A.9):

$$A + B = \begin{bmatrix} 1 & 4.5 & 4 \\ 3 & 3.5 & 4 \\ 2 & 4.5 & 0.5 \end{bmatrix} \tag{A.9}$$

A.1.6 MATRIX MULTIPLICATION

We can multiply two matrices if the first matrix has the same number of columns as the second has rows. In other words, if we have an (m_1, n_1) matrix A and a (m_2, n_2) matrix B then we can only multiply them if $n_1 = m_2$. We begin by considering the matrix B as n_2 vectors, with each vector in $\mathbb{R}^{m_2}$. Eq. (A.10) illustrates this:

$$\left[\begin{bmatrix} b_{1,1} \\ b_{2,1} \\ \vdots \\ b_{n_2,1} \end{bmatrix} \begin{bmatrix} b_{1,2} \\ b_{2,2} \\ \vdots \\ b_{n_2,2} \end{bmatrix} \cdots \begin{bmatrix} b_{1,m_2} \\ b_{2,m_2} \\ \vdots \\ b_{n_2,m_2} \end{bmatrix}\right] \tag{A.10}$$

We treat each column in the second matrix as a column vector, so we can multiply the first matrix by ea column as shown in Eq. (A.7) and create a new vector. The new vectors have the same number of rows as the first matrix and the same number of columns as the second matrix. This means if we multiply a (n_1, m_1) matrix by a (n_2, m_2) matrix, the result is a (n_1, m_2) matrix.

Example A.1.5. We begin with two matrices A and B in Eq. (A.11):

$$A = \begin{bmatrix} 0 & 2 & 3 \\ 1 & 4.5 & 2.5 \end{bmatrix} \quad B = \begin{bmatrix} 1 & 5 & 1 \\ 2 & 3.5 & 3 \\ 0.5 & 3 & 0 \end{bmatrix} \tag{A.11}$$

The element in the first row and column of the matrix multiplication is given by $0 \times 1 + 2 \times 2 + 3 \times (0.5) = 5.5$. Eq. (A.12) is the matrix AB:

$$AB = \begin{bmatrix} 5.5 & 16 & 6 \\ 10 & 28.25 & 14.5 \end{bmatrix} \tag{A.12}$$

Let X be a square matrix with n rows and columns. The **identity** matrix is a square matrix I with n rows and columns such that $XI = X$. The identity matrix is given by 1's on the diagonal of the 0's everywhere else as illustrated in Eq. (A.13):

$$I = \begin{bmatrix} 1 & 0 & 0 & \cdots & 0 \\ 0 & 1 & 0 & \cdots & 0 \\ 0 & 0 & 1 & \cdots & 0 \\ & & & \cdots & 0 \\ 0 & 0 & 0 & \cdots & 1 \end{bmatrix} \tag{A.13}$$

If A is an (n, n) matrix, then we can multiply it by itself and get an (n, n) matrix as the result. The multiplication is usually denoted by A^2. We can continue this process, that is, multiply A^2 by A and get A^3. This process is called **iteration**.

A.2 EIGENVALUES

A **fixed point** of a function $f : A \to A$ is an element $a \in A$ such that $f(a) = a$. Since we know a matrix is a linear mapping on the space of vectors, then let M be an (n, m) matrix. A vector $\bar{v} \in \mathbb{R}^m$ is a fixed point of M if $Mv = v$. We also know that a linear mapping respects scalar multiplication.

We will define a pair $\lambda \in \mathbb{R}$ and $\bar{x} \in \mathbb{R}^m$ an **eigenvalue** and **eigenvector** pair if Eq. (A.14) holds:

$$M(\bar{x}) = \lambda\bar{x} \tag{A.14}$$

If there are n columns in a matrix, there are at most n eigenvector and eigenvalue pairs.

A.3 ADDITIONAL MATRIX OPERATIONS

The **transpose** of a matrix is a reflection of the matrix along the diagonal of the matrix. In other words, an element in the ith row and jth column becomes an element of the jth row and ith column. If A is a (m, n) matrix, then the transpose becomes a (n, m) matrix. We also represent the transpose of A by A^T.

If a matrix U is such that $U = U^T$, then it is called **symmetric**.

Let A be a square matrix with n rows and columns. If there exists a matrix B such that $AB = I$, then B is called the **inverse** matrix. B is commonly denoted by A^{-1}.

Another operation on a square matrix is called **trace**. Let A our (n, n) matrix. The trace of A, denoted $tr(A)$, is the sum of the diagonal elements of the matrix. Eq. (A.15) is the explicit definition of trace:

$$tr(A) = \sum_{i=1}^{n} a_{i,i} \tag{A.15}$$

Bibliography

Agah, A., Das, S.K., Basu, K., Asadi, M., 2004. Intrusion detection in sensor networks: a non-cooperative game approach. In: Proceedings of the Third IEEE International Symposium on Network Computing and Applications (NCA 2004). IEEE, pp. 343–346.

Akkiraju, N., Edelsbrunner, H., Facello, M., Fu, P., Mücke, E., Varela, C., 1995. Alpha shapes: definition and software. In: Proceedings of the First International Computational Geometry Software Workshop, pp. 63–66.

Axelrod, R., 1997. The evolution of strategies in the iterated prisoner's dilemma. In: The Dynamics of Norms. Cambridge University Press, Cambridge, pp. 1–16.

Baldi, P., Frasconi, P., Smyth, P., 2003. Modeling the Internet and the Web: Probabilistic Methods and Algorithms. Wiley, Chichester.

Basu, A., Ong, C.H.L., Rasala, A., Shepherd, F.B., Wilfong, G., 2002. Route oscillations in I-BGP with route reflection. ACM SIGCOMM Comput. Commun. Rev. 32 (4), 235–247.

Beezer, R.A., 2008. A First Course in Linear Algebra.

Bondy, J.A., Murty, U.S.R., 1976. Graph Theory With Applications, vol. 290. Macmillan, London.

Calvert, K.L., Doar, M.B., Zegura, E.W., 1997. Modeling Internet topology. IEEE Commun. Mag. 35 (6), 160–163.

Casey, W., Morales, J.A., Nguyen, T., Spring, J., Weaver, R., Wright, E., Metcalf, L., Mishra, B., 2014. Cyber security via signaling games: toward a science of cyber security. In: Distributed Computing and Internet Technology. Springer, New York, pp. 34–42.

Chakrabarty, K., Iyengar, S.S., Qi, H., Cho, E., 2002. Grid coverage for surveillance and target location in distributed sensor networks. IEEE Trans. Comput. 51 (12), 1448–1453.

Cranor, C.D., Gansner, E., Krishnamurthy, B., Spatscheck, O., 2001. Characterizing large DNS traces using graphs. In: Proceedings of the First ACM SIGCOMM Workshop on Internet Measurement. ACM, pp. 55–67.

Dawood, H.A., 2014. Graph theory and cyber security. In: Proceedings of the Third International Conference on Advanced Computer Science Applications and Technologies (ACSAT). IEEE, pp. 90–96.

Deza, M.M., Deza, E., 2009. Encyclopedia of Distances. Springer, New York.

Diestel, R., 2000. Graph theory. In: Graduate Texts in Mathematics, vol. 173. Springer-Verlag GmbH, Berlin.

Doebeli, M., Hauert, C., 2005. Models of cooperation based on the prisoner's dilemma and the snowdrift game. Ecol. Lett. 8 (7), 748–766.

Edelsbrunner, H., Harer, J., 2008. Persistent homology—a survey. Contemp. Math. 453, 257–282.

Edelsbrunner, H., Harer, J., 2010. Computational Topology: An Introduction. American Mathematical Society, London.

Edelsbrunner, H., Letscher, D., Zomorodian, A., 2002. Topological persistence and simplification. Discrete Comput. Geom. 28 (4), 511–533.

Faghani, M.R., Nguyen, U.T., 2013. A study of XSS worm propagation and detection mechanisms in online social networks. IEEE Trans. Inf. Forensics Security 8 (11), 1815–1826.

Freeman, L.C., 1977. A set of measures of centrality based on betweenness. Sociometry 35–41.

Ghrist, R., 2008. Barcodes: the persistent topology of data. Bull. Am. Math. Soc. 45 (1), 61–75.

Grinstead, C.M., Snell, J.L., 2012. Introduction to Probability. American Mathematical Society, London.

Harary, F., 1969. Graph Theory. Addison-Wesley, Reading, MA.

Hungerford, T., 2012. Abstract Algebra: An Introduction. Cengage Learning, Boston, MA.

Iliofotou, M., Pappu, P., Faloutsos, M., Mitzenmacher, M., Singh, S., Varghese, G., 2007. Network monitoring using traffic dispersion graphs (TDGS). In: Proceedings of the Seventh ACM SIGCOMM Conference on Internet Measurement. ACM, pp. 315–320.

Irwin, B., Pilkington, N., 2008. High level Internet scale traffic visualization using Hilbert curve mapping. In: VizSEC 2007. Springer, New York, pp. 147–158.

Janies, J., 2008. Existence plots: a low-resolution time series for port behavior analysis. In: Visualization for Computer Security. Springer, New York, pp. 161–168.

Jiang, N., Cao, J., Jin, Y., Li, L.E., Zhang, Z.L., 2010. Identifying suspicious activities through DNS failure graph analysis. In: Proceedings of the 18th IEEE International Conference on Network Protocols (ICNP). IEEE, pp. 144–153.

Johansson, S., Carlsson, B., Boman, M., 1999. Modeling strategies as generous and greedy in prisoner's dilemma like games. In: Simulated Evolution and Learning. Springer, New York, pp. 285–292.

Jolliffe, I., 2002. Principal Component Analysis. Wiley Online Library.

Jost, J., 2013. Game Theory. Mathematical and Conceptual Aspects.

Konishi, S., Kitagawa, G., 2008. Information Criteria and Statistical Modeling. Springer Science & Business Media, New York.

Kraaikamp, F.D.C., Meester, H.L.L., 2005. A Modern Introduction to Probability and Statistics, Springer-Verlag, London.

Leyton-Brown, K., Shoham, Y., 2008. Essentials of game theory: a concise multidisciplinary introduction. Syn. Lect. Artif. Intell. Mach. Learn. 2 (1), 1–88.

Liu, C., Albitz, P., 2006. DNS and Bind. O'Reilly Media, Inc, Sebastopol, CA.

Lock, R.H., Lock, P.F., Morgan, K.L., 2012. Statistics: Unlocking the Power of Data. Wiley Global Education, Hoboken, NJ.

Manning, C.D., Schütze, H., 1999. Foundations of Statistical Natural Language Processing. MIT Press, Cambridge, MA.

Mazalov, V., 2014. Mathematical Game Theory and Applications. John Wiley & Sons, Hoboken, NJ.

Moscibroda, T., O'Dell, R., Wattenhofer, M., Wattenhofer, R., 2004. Virtual coordinates for ad hoc and sensor networks. In: Proceedings of the 2004 Joint Workshop on Foundations of Mobile Computing. ACM, pp. 8–16.

Munkres, J.R., 1975. Topology: A First Course. Prentice-Hall, Englewood Cliffs, NJ.

Munkres, J.R., 1984. Elements of Algebraic Topology, Addison-Wesley, Menlo Park, CA.

Nagaraja, S., Mittal, P., Hong, C.Y., Caesar, M., Borisov, N., 2010. BotGrep: finding P2P bots with structured graph analysis. In: USENIX Security Symposium, pp. 95–110.

Nisan, N., Roughgarden, T., Tardos, E., Vazirani, V.V., 2007. Algorithmic Game Theory, vol. 1. Cambridge University Press, Cambridge, UK.

Osborne, M.J., Rubinstein, A., 1994. A Course in Game Theory. MIT Press, Cambridge, MA.

Plummer, M.D., 1970. Some covering concepts in graphs. J. Comb. Theory 8 (1), 91–98.

Raga'ad, M.T., Keller, P., Ebert, A., Garth, C., Middel, A., Hagen, H., 2011. A general introduction to graph visualization techniques. In: VLUDS, pp. 151–164.

Ramsden, A., Bate, A., 2008. Using Word Clouds in Teaching and Learning. Other. University of Bath.

Rapoport, A., Chammah, A.M., 1965. Prisoner's Dilemma: A Study in Conflict and Cooperation, vol. 165. University of Michigan Press, Michigan.

Rekhter, Y., Li, T., Hares, S., 2005. A border gateway protocol 4 (BGP-4). Technical Report.

Riaz, F., Ali, K.M., 2011. Applications of graph theory in computer science. In: Proceedings of the Third International Conference on Computational Intelligence, Communication Systems and Networks (CICSyN). IEEE, pp. 142–145.

Roy, S., Ellis, C., Shiva, S., Dasgupta, D., Shandilya, V., Wu, Q., 2010. A survey of game theory as applied to network security. In: Proceedings of the 43rd Hawaii International Conference on System Sciences (HICSS). IEEE, pp. 1–10.

Sammon, J.W., 1969. A nonlinear mapping for data structure analysis. IEEE Trans. Comput. (5), 401–409.

Saxe, J., Mentis, D., Greamo, C., 2012. Visualization of shared system call sequence relationships in large malware corpora. In: Proceedings of the Ninth International Symposium on Visualization for Cyber Security. ACM, pp. 33–40.

Schwarzlander, H., 2011. Probability Concepts and Theory for Engineers. John Wiley & Sons, Hoboken, NJ.

Shiva, S., Roy, S., Dasgupta, D., 2010. Game theory for cyber security. In: Proceedings of the Sixth Annual Workshop on Cyber Security and Information Intelligence Research. ACM, p. 34.

Shubik, M., Weber, R.J., 1981. Systems defense games: colonel blotto, command and control. Nav. Res. Logist. Q. 28 (2), 281–287.

Sieradski, A.J., 1992. An Introduction to Topology and Homotopy. Wadsworth Pub. Co, Boston, MA

Skyrms, B., 2001. The stag hunt. In: Proceedings and Addresses of the American Philosophical Association. JSTOR, pp. 31–41.

Solomon, F., 1987. Probability and Stochastic Processes. Prentice Hall, Englewood Cliffs, NJ.

Spiegelhalter, D., Best, N., Carlin, B., van der Linde, A., 2002. Bayesian measures of model complexity and fit. J. R. Stat. Soc. Ser. B 64 (4), 583–639.

Stolfo, S.J., Hershkop, S., Wang, K., Nimeskern, O., Hu, C.W., 2003. A behavior-based approach to securing email systems. In: Computer Network Security. Springer, New York, pp. 57–81.

Sun, K., Peng, P., Ning, P., Wang, C., 2006. Secure distributed cluster formation in wireless sensor networks. In: Proceedings of the 22nd Annual Computer Security Applications Conference (ACSAC'06). IEEE, pp. 131–140.

Thie, P.R., Keough, G.E., 2011. An Introduction to Linear Programming and Game Theory. John Wiley & Sons, New York.

Thomas, M., Metcalf, L., Spring, J., Krystosek, P., Prevost, K., 2014. Silk: a tool suite for unsampled network flow analysis at scale. In: Proceedings of the 2014 IEEE International Congress on Big Data (BigData Congress). IEEE, pp. 184–191.

Turán, P., 1941. On an extremal problem in graph theory. Mat. Fiz. Lapok 48 (436–452), 137.

Walker, B., 2008. Using persistent homology to recover spatial information from encounter traces. In: Proceedings of the Ninth ACM International Symposium on Mobile Ad Hoc Networking and Computing. ACM, pp. 371–380.

Wattenhofer, R., Zollinger, A., 2004. XTC: a practical topology control algorithm for ad-hoc networks. In: Proceedings of the 18th International Parallel and Distributed Processing Symposium. IEEE, p. 216.

Watts, D.J., Strogatz, S.H., 1998. Collective dynamics of small-world networks. Nature 393 (6684), 440–442.

Wilkinson, L., Friendly, M., 2009. The history of the cluster heat map. Am. Stat. 63 (2).

Yavvari, C., Tokhtabayev, A., Rangwala, H., Stavrou, A., 2012. Malware characterization using behavioral components. In: Computer Network Security. Springer, New York, pp. 226–239.

Zhu, Z., Cao, G., Zhu, S., Ranjan, S., Nucci, A., 2012. A social network based patching scheme for worm containment in cellular networks. In: Handbook of Optimization in Complex Networks. Springer, New York, pp. 505–533.

Zomorodian, A.J., 2005. Topology for Computing, vol. 16. Cambridge University Press, Cambridge, UK.

Index

Note: Page numbers followed by *f* indicate figures and *t* indicate tables.

R

S

Printed in the United States
By Bookmasters